Tunable External Cavity Diode Lasers

Tunable External Cavity Diode Lasers

Cunyun Ye
Texas A & M University, USA

NEW JERSEY • LONDON • SINGAPORE • BEIJING • SHANGHAI • HONG KONG • TAIPEI • CHENNAI

Published by

World Scientific Publishing Co. Pte. Ltd.
5 Toh Tuck Link, Singapore 596224
USA office: 27 Warren Street, Suite 401–402, Hackensack, NJ 07601
UK office: 57 Shelton Street, Covent Garden, London WC2H 9HE

British Library Cataloguing-in-Publication Data
A catalogue record for this book is available from the British Library.

TUNABLE EXTERNAL CAVITY DIODE LASERS
Tunable Semiconductor Diode Lasers

ISBN-13 978-981-256-088-9
ISBN-10 981-256-088-2

Editor: Tjan Kwang Wei

Printed in Singapore

To my wife and daughters

Jane Qin Chen and Christina Ye, Vivian Ye

Preface

Since its invention in 1961, the semiconductor diode laser has been the most important of all types of lasers. It has found widespread applications in numerous fields ranging from basic research to industrial photonic systems such as coherent fiber telecommunications because of its well-known features: mass productivity, high reliability, miniature size, lower power consumption, wide tunability, high efficiency, and excellent direct modulation capability.

There are numerous books on semiconductor diode lasers and tunable lasers. However, to the best of my knowledge, this is the first book on tunable external cavity semiconductor diode lasers that provides an up-to-date exploration on physics, technology, and performance of widely applicable coherent radiation sources of tunable external cavity diode lasers. I hope this book give undergraduate and graduate students, scientists, and engineers the practical information needed to study the tunable external cavity diode laser; to build up the systems of the tunable external cavity diode laser, and to develop advanced laser systems for their own particular applications.

The purpose of this book is to provide a thorough account of the state-of-the-art in relation to tunable external cavity diode lasers. This is accomplished by explaining this account with basic concepts of semiconductor diode laser and its tunability with monolithic structures as described in Chapters 2 and 3, characteristic features of components and system of tunable external cavity diode lasers as elucidated in Chapters 4, 5 and 6, frequency stabilization of external cavity diode lasers introduced in Chapters 7 and applications of external cavity diode lasers in a wide variety of fields investigated in chapter 8. In this way, this book can be used in a variety of ways by a broad range of readers: as a text book in special topics for undergraduate and graduate students; as a handbook or reference book for researchers, scientists and engineers who are working in related areas.

This book starts with the fundamentals of semiconductor diode lasers, including principles of diode laser operation as well as diode laser structure and its spectral characteristics. Then the essential properties of single-mode tunable monolithic laser diodes are summarized. The principles, design, and practice of tunable external cavity diode lasers are described with emphasis on their wide tuning, narrow linewidth, high power, and extreme stability. Stabilization of external cavity diode lasers by various methods is briefly introduced. Numerous examples of applications of tunable external cavity diode lasers in different fields are developed. A brief outlook concludes this book. The material contained in this book are representative of state-of-the-art research and development in tunable external cavity diode lasers. I hope the readers will find this book useful.

I would like to take this opportunity to thank many people who gave so generously of their time to help me finish this book. First of all, Dr. Marlan O. Scully was not only my Ph.D. advisor for my dissertation on laser physics and quantum optics but also illuminated science for me as well as offered me the opportunity to do postdoctoral research at Texas A & M University. Without his support, this book would not exist. Secondly, I am deeply indebted to Dr. Shiyao Zhu, who brought me to the family of Dr. Scully, the best in the world, to study laser physics and quantum optics. Thirdly, I acknowledge the help from all my colleagues with this book, including Dr. Alexander S. Zibrov, Dr. Zuhail S. Zubairy, Dr. Yuri V. Rostovtsev, Dr. George R. Welch, Dr. P. R. Hemmer, Dr. Alexy Beleinyn, Dr. Edward S. Fry, Dr. Kishor T. Kapale, and special thanks to Kim Chapin and Terri Tomaszek.

Finally, I want to acknowledge the tireless efforts of the editor and editorial staff members for making sure that this book was produced.

C. Y. Ye

Contents

Chapter 1

Introduction

In this introduction, we demonstrate the reasons why we need tunable external cavity diode lasers by giving some typical examples of their widespread applications in various areas. A brief history of semiconductor lasers is then given. An overview of tunable monolithic semiconductor lasers and tunable external cavity diode lasers is introduced. A comparison of different tunable diode laser technologies concludes this introduction.

1.1 Need for tunable external cavity diode lasers

Tunable external cavity diode lasers are of considerable interests in coherent optical telecommunications, atomic and molecular laser spectroscopy, precise measurements, and environmental monitors. With the development of semiconductor diode lasers, tunable external cavity diode lasers are finding a wide variety of applications in a broad range of fields. An overview of the most important applications is outlined as follows:

- Optical coherent telecommunications
  - (1) Reduction of inventory and sparing
  - (2) Easy access for new service without hardware change
  - (3) Drop-add multiplexers
  - (4) Elimination of wavelength blocking
  - (5) Easy use of optical core
- Sensing
  - (1) Ultra-high resolution spectroscopy
  - (2) Optical radar
  - (3) Atmospheric and environmental studies

(4) Industrial processing monitoring

- Precise measurements

(1) Atomic clock and magnetometer
(2) Optical spectrum analysis
(3) Device characterization

In addition to typical applications exemplified above, there are many other applications, including nonlinear optical conversion, optical data storage, coherent optical transient processing, and quantum optical manipulation and engineering.

1.2 Brief introduction of semiconductor diode lasers

Semiconductor diode lasers, since it was first demonstrated in several laboratory more than forty years ago [Bernard and Duraffoug (1961); Basov *et. al.* (1961); Hall *et. al.* (1962); Nathan (1962); Holonyak and Bevacqua (1962)], are thought to be an innovation that has revolutionized the world in which we are living. As early as the late 1960s and early 1970s, semiconductor lasers were found the utility in optical data storage, fiber optic, and free space communications. However in its early stage, the simple p-n homojunction device was far away from the practical use because it needs large injection current and at cryogenic temperature. The development of double heterostructure lasers [Hayashi *et. al.* (1970); Alferov *et. al.* (1970)] and subsequently the breakthroughs in device design made possible to fabricate reliable diode lasers that operate with sufficiently low currents at room temperature [Thompson (1980)]. In many points of view, semiconductor lasers are second only to the transistor and integrated circuit as to their influences on modern technological arena .

However, advances in diode laser designs strongly depend on the fabrication technologies. Liquid phase epitaxy (LPE) is used to fabricate simple configuration of diode lasers, the performance is limited by the inability of LPE to grow uniform thin epitaxial layers and accurately tailored doping profiles that result in low power output of a diode laser. The first important technology improvement necessary for the realization of high-power lasers was the development of two new growth technologies: metal organic chemical vapor deposition (MOCVD) and molecular beam epitaxy (MBE). These two key technologies produced a powerful tool that enables the laser designer to control the crystal deposition to an atomic scale, which lead to uniform material deposition and ultimately quantum well active layers.

With these technologies, the next development is in the 1980s and the early 1990s. More complicated semiconductor devices were developed including single-mode lasers with high modulation speed, diode laser arrays with high power output, distributed feedback lasers for use in a long-haul optical communication system[Green (1993); Agrawal (1992)]. Moreover, progresses in engineering new diode laser material covering emission wavelength from the violet-blue to mid- and far-infrared have been motivated by replacing many kinds of bulk gas and solid state lasers with compact, lost cost, and efficient semiconductor lasers. The utilization of quantum well heterostructure, especially strained structures, makes it possible to operate quantum well diode lasers at very low threshold current (sub-mA) with higher efficiencies and high modulation speed (multi-GHz) [Buus (1991); Agrawal and Dutta (1993); Zory (1993); Carroll *et. al.* (1998)].

The development of cavity configurations with state-of-the-art epitaxial technology have recently attained extremely narrow linewidth(kHz) and wide tuning ranges for diode lasers. Laser structures incorporating micro and nano-structure such as micro- cavity, quantum wires, and quantum dots have demonstrated the improved lower noise, ultrahigh speed modulation and even high conversion efficiency. Quantum cascade lasers and blue-violet lasers have been widely used in many applications with better understanding of physical principles of semiconductor lasers[Nakamura and Fasol (1997); Kapon (1999)].

1.3 Review of tunable diode lasers

Tunable diode laser are developed primarily for applications in wavelength division multiplexing (WDM) technology in coherent communication systems, sensing, as well as in precise measurements[Wieman and Hollberg (1991); Yamamoto (1991); Fox *et. al.* (1997)].

1.3.1 *Tunable monolithic semiconductor lasers*

Wavelength tunable lasers are very desirable for increasing the capacity, functionality, specialty, and flexibility of the current and next generation all-optical devices and networks. An ideal continuously tunable single mode laser is one that the output can be tuned smoothly over the whole gain bandwidth without any significant reduction in output power.

The important factors for tunable diode lasers include the tuning range and speed, side-mode suppression ratio (SMSR), spectral purity, output power and reliability [Kobayashi and Mito (1988); Koch and Koren (1990) ; Amann and Buss (1998); Coldren (2000)]. There are at least four various configurations of tunable lasers in the applications of telecommunication networks and other fields. All of these tunable lasers have promising to replace the current commonly-used distributed feedback (DFB) [Kogelnik and Shank (1971)] lasers.

One of the DFB configurations can achieve their wavelength tuning by thermal control over a small range of 2 nm to 4 nm with a high power output of more than 20 mW. Improvements of DFB configuration can lead to the widely tuning range to 10 nm to 15 nm with reduced output power.

The next configuration is the distributed Bragg reflector (DBR) [Wang (1974)] lasers that use a gain medium sandwiched between two Bragg grating sections whose optical properties are changed by current injection. Different modifications of the DBR include the super-structure grating DBR (SSG-DBR), the sampled-grating DBR (SG-DBR), and the grating-assistant codirectional coupler (GACC) DBR. Tunable DBR usually present a wide tuning range of about 40 nm, however, output power is reduced by the extensive losses in the tuning sections.

Another configuration is a vertical-cavity surface-emitting laser (VCSEL) [Chang-Hasnain (2000); Li and Iga (2003)] with a MEMS based tuning element. Typically, the top mirror of the VCSEL is mechanically supported by one or multiple spring. An electrostatic potential actuates the mirror resulting in the tuning of the cavity length, and hence its lasing wavelength, the tuning range can be up to 40 nm. The output power depends on whether the VCSEL is electrically or optically pumped.

The laser arrays DFB have been demonstrated both with monolithic waveguide combiner as well as with external optics. MEMS mirror has been used to select one of the several lasers to couple to a single output fiber [Berger and Anthon (2003)].

1.3.2 *Tunable external cavity diode lasers*

There are many possible techniques for the wavelength tunability of semiconductor diode lasers as describe above. One of practical approaches is tunable external cavity diode lasers (ECDLs), which could provide an alternative to monolithic semiconductor diode laser for accomplishing the widely tuning of diode laser. Tunable external cavity diode laser system consists primarily of a semiconductor diode laser with or without antireflection coatings on one or two facets, collimator for coupling the output of a diode laser, and an external mode-selection filter. In general, the features of a diode laser in an external cavity can change greatly, depending on the length of external cavity, feedback strength, optical power, and diode laser parameters. Littman-Metcalf and Littrow cavity configurations are typical examples of ECDLs source in which gratings are used to provide optical feedback, select single-mode operation, and tune the wavelength over the whole range of gain bandwidth by moving and rotating the grating position [Duarte (1996); Zorabedian (1996)].

1.3.3 *Comparisons of technologies*

After having discussed the various tunable diode laser, we summarize the key performance properties of these technologies in Table 1.1. The column of tuning range illustrates wavelength tuning range achievable. Tunable external cavity diode lasers show the widest tuning range. Multiple-section distributed-Bragg-reflector (DBR) can cover a quite wide range of wavelength. Distributed feedback (DFB) is rather limited due to its inherent wavelength stability, but its arrayed devices have wide tuning ranges. The VCSEL can in principle have a broad tuning range.

Table 1.1 Comparison of tunable diode laser technologies.

Devices	Tuning range	Spectral purity	Mode control	Tuning speed	Modulation speed
DBR	8 ∼80 nm	>40 dB	Fair	Fast(μs)	Fast(GHz)
DFB array	3∼ 4 $nm\times$ # of DFBs	>55 dB	Very good	Slow(ms)	Fast(multi-GHz)
VCSEL	>12 nm	∼45 dB	Hard	Fast(μs)	Slow(<GHz)
ECDL	>32 nm	>50 dB	Fair	slow(ms)	Slow(<GHz)

Spectral purity includes both linewidth of diode laser and side-mode suppression ratio (SMSR). Tunable external cavity diode laser behaves itself good spectral characteristics. DFB laser is famous for its very narrow line width with good SMSR. VCSEL provide good spectral performance due to the restriction of spurious longitudinal modes. DBR spectral characteristic is somehow inferior to other devices since side mode suppression is difficult to maintain. However, tunable DBR lasers have demonstrated sufficiently narrow linewidth for most applications in communications.

Modulation speed refers to the ability to directly modulate the laser diode at tens of GHz speed while maintaining good spectral performance. DFB have nice high speed modulation behavior. Therefore, DFB array are used to direct high speed modulation in telecommunications. Tunable DBR is not good enough for high speed modulation because of tuning complexities. External cavity diode laser have poor modulation bandwidths, VCSEL are being a candidate for high speed modulation.

The majority of these tunable lasers suffer from slow tuning speed. DFB array requires a temperature change. Whereas external cavity and tunable VCSEL need a mechanical moving. Fortunately, small DFB array thermal change and small variation of mechanical system of VCSEL allow one to have millisecond tuning speed. Tunable DBR can be very rapidly tuned with sub-nanosecond time scale.

The other questions concerned with tunable diode laser are their output power, which is quite similar for all of these tunable diode lasers. The power can be greatly amplified by semiconductor optical amplifiers(SOA).

Chapter 2

Basics of Semiconductor Diode Lasers

The development of the semiconductor laser diode traces its origins to the early 1960s shortly after the invention of other laser systems. Semiconductor diode laser has been the most important of all lasers since it was invented in 1961 [Quist *et. al.* (1962)]. Semiconductor diode lasers are used in a wide variety of applications ranging from readout sources in compact disk players to transmitters in optical fiber communication systems because they have well-known features: high reliability, miniature size, lower power consumption, wide tunability, high efficiency, and excellent direct modulation capability, and other characteristics.

In this chapter, we shall introduce some of the basic ingredients needed to understand the operation of semiconductor diode lasers [Kittle (1982); Yariv (1991); Coldren and Corzine (1995)]. First, the interaction of light with a two energy level system and the requirements for lasing in the semiconductor laser are introduced. Then energy levels and bands in various semiconductor laser structures are described. Finally basic characteristics of semiconductor diode lasers are briefly reviewed.

2.1 Principle of diode laser operation

In this section we explore the fundamentals of light absorption, spontaneous emission, and stimulated emission induced in a two-level system in a single atom or molecule with a monochromatic electromagnetic wave. We also describe the basic conditions necessary for laser operation in a semiconductor laser [Svelto (1998)].

2.1.1 *Absorption, spontaneous emission, and stimulated emission*

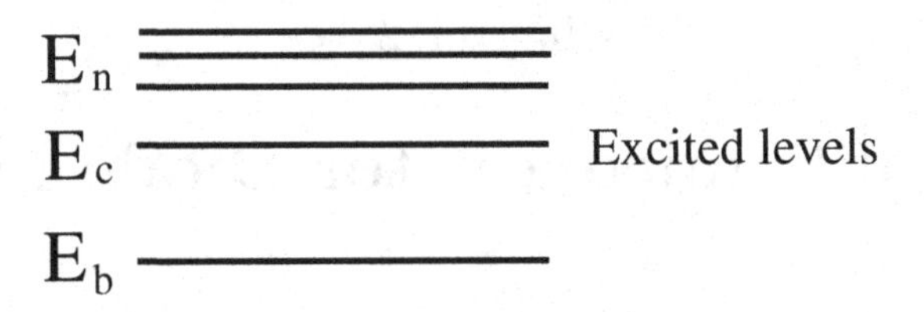

Fig. 2.1 Energy level diagram.

Any electron in an atom or molecule has its own stable orbits that is called stationary states, in which atom has a specific energy level as shown in Fig. 2.1. Atom radiates in terms of light emission when an electron makes a transition from one stationary state to another. Frequency of the radiation is related to the energies of the orbits by Bohr's principle

$$\nu = \frac{E_f - E_i}{h}, \tag{2.1}$$

where E_f, E_i are any energy levels of final and initial states in an atom or molecule, h=6.625×10^{-34}Joul·sec is Planck's constant. There are three different kinds of electron transitions between two different states by the interaction with the light, the processes are shown in Fig. 2.2. The first type of transition of an atom illustrated in Fig. 2.2(a) is referred to as resonant absorption. Assume an atom is initially residing in the ground state with energy E_i, the atom stays in the ground state until light with its frequency $\nu = \nu_0$ is applied, where ν_0 is the transition frequency between two energy levels. In this case there is a most probability for an atom to make transition from lower E_i to the higher level E_f by absorption of a quanta of light. This is called resonant absorption process.

Fig. 2.2(b) presents spontaneous emission. When an electron jumps

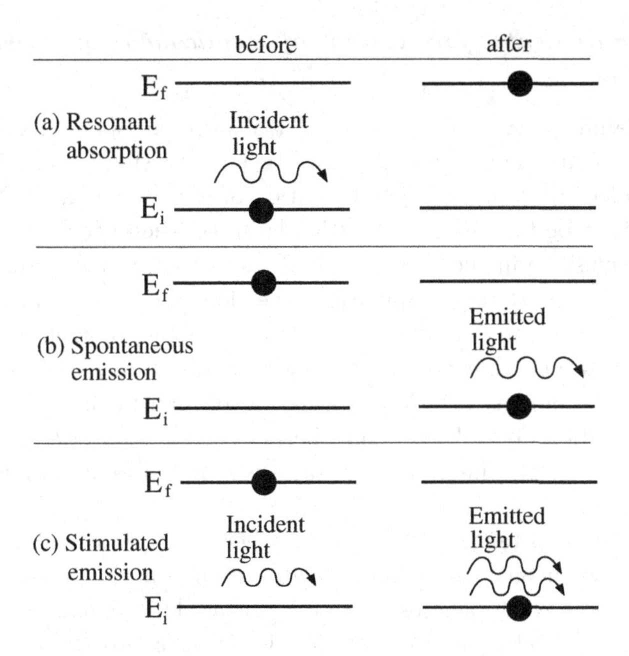

Fig. 2.2 Three fundamental radiation processes associated with the interaction of light with an atom or molecule. (a) resonant absorption, (b) spontaneous emission, and (c) stimulated emission.

to the higher energy level by absorbing light, an atom has a tendency to decay to the lower stable level with a definite lifetime in the final level, the corresponding energy difference $E_f - E_i$ must be therefore released by spontaneously emitting a quanta of light. Since each electron makes a transition independently, light is emitted in all direction at random phase, such light is incoherent as compared to coherent light later on we will see, this process is defined as spontaneous emission phenomenon.

The third kind of transition shown in Fig. 2.2(c) is referred to as stimulated emission. Suppose an electron is initially found in the final state and light with frequency $\nu = \nu_0$ is incident in an atom or molecule, the incident light will enforce an atom to undergo transition from $\mathrm{E}_f \rightarrow \mathrm{E}_i$ in such a way that a new light is generated in addition to the incident light, this phenomenon is called stimulated emission. The generated light has the same phase and direction as that of incident light, such stimulated emission light is known as coherent light.

2.1.2 *Requirements for lasing of semiconductor diode lasers*

We have examined basic concepts of light absorption, spontaneous emission, and stimulated emission. In order to generate lasing, one must make atoms or molecules working in the condition of stimulated emission. Two conditions must be fulfilled to have stimulated emission occurred: Firstly, more atoms must be in the higher excited states than in the lower energy levels, i.e., there must be a population inversion (necessary condition for lasing). This is necessary otherwise stimulated emissions of light be directly re-absorbed by atoms that populate in lower energy states. But usually an atom has more population in lower energy level than in the higher levels because population obeys Boltzmann thermal distribution law. Therefore, external pumping of atoms to higher states is required to accomplish population inversion.

The second important condition is that there be more stimulated emissions than spontaneous ones in an active medium (sufficient condition for lasing). Therefore, an optical resonator must be used to feedback new generated coherent light into the medium. The two most important components in a laser are thus active light-emitting medium and optical resonator for regenerating the radiation field besides the pumping.

As atoms and molecules are squeezed into semiconductor crystal, the discrete atomic energy level smears into energy bands of solid, which is significantly different from discrete energy levels, as shown in Fig. 2.3. Semiconductor valence band is formed by multiple splitting of the highest occupied atomic energy level of the constituent atoms; Likewise, the next higher-lying atomic level splits apart into conduction band.

Consider some electrons in the valance band are excited to the conduction band via proper pumping scheme for nondegenerate semiconductor. Initially the conduction band is completely empty, to make things simple and without loss of generosity, we start off our analysis by supposing the absolute temperature to be T=0 K. The valence band is filled with electrons while the conduction band is empty as indicted in Fig. 2.4(a). In between these bands is the forbidden region of the bandgap, the energy difference between the bottom of conduction band and the top of the valence band is the bandgap energy E_g whose value differs for different materials. The valence band is completely filled, if some electrons are excited from valence band to conduction band by forward bias current, after about 1 ps short time, electrons in the conduction band spontaneously drop to the lowest

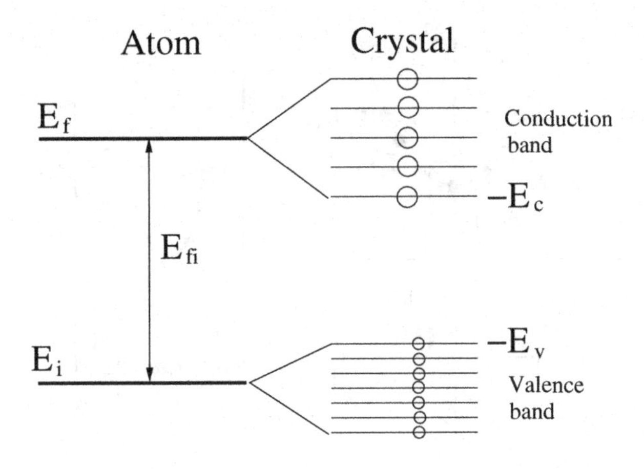

Fig. 2.3 Energy level diagram for atom and solids.

unoccupied levels in the intraband. We name the upper boundary of the electron energy levels in the conduction band the quasi-Fermi level E_{Fc}. The occupancy probability of electron in the quasi-Fermi level, as shown in Fig. 2.4(b), is

$$f_c = \frac{1}{1 + e^{(E_2 - E_{Fc})/kT}}, \tag{2.2}$$

where E_2 is a given level of energy in the conduction band, k is Boltzmann constant. Meanwhile holes appear in the valence band when electrons near the top of the valence band drop to the lowest energy levels of the unoccupied valence energy levels, leaving on the top of the valence band an empty part. We call the new upper boundary energy level of the valence band quasi-Fermi level E_{Fv}. The occupancy probability of holes in the quasi-Fermi level is

$$f_v = \frac{1}{1 + e^{(E_1 - E_{Fv})/kT}}, \tag{2.3}$$

where E_1 is a certain level of energy in the valence band. At T=0, f(E)=1 for $E < E_F$, and f(E)=0 for $E > E_F$ so that all levels below Fermi level are occupied while those above it are empty.

When electrons in the conduction band run into the valence band, they will combine with the holes, in the same time they generate photons, this is recombination radiation. In order to make this recombination radiation to lase, several conditions must be met, one can see these prerequisites

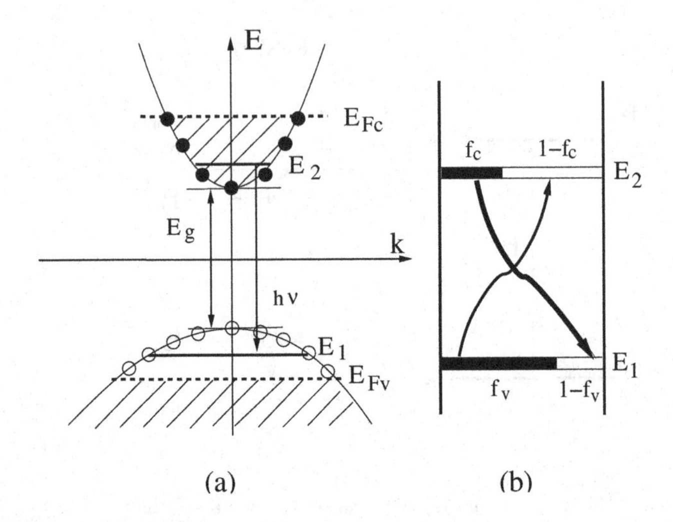

Fig. 2.4 (a) Electron energy versus wave number for a bulk semiconductor, and transitions between energy levels in the conduction and valence bands. (b) Level occupation probability f_c(E) and f_v(E) for conduction and valence bands under thermal equilibrium within each band. The finite width of energy levels indicate the finite energy uncertainty due to relaxation processes limiting carrier life time.

from band to band transitions. Fig. 2.5 illustrates band to band stimulated absorption and emission process, the rate of stimulated absorption is given by

$$R_{12} = R_r[f_v(1 - f_c)], \tag{2.4}$$

which outlines photon absorption that stimulates generation of an electron in the conduction band while leaving a hole in the valence band. Where R_r is the transition rate of two levels in the respective band. The rate of stimulated emission is given by

$$R_{21} = R_r[f_c(1 - f_v)], \tag{2.5}$$

which shows that an incident photon interacts with a system, stimulating combination of an electron and a hole, and simultaneously generating a new photon. The net stimulated emission rate is

$$R_{net} = R_{21} - R_{12} = R_r(f_c - f_v). \tag{2.6}$$

This result is of central importance to understand population inversion in semiconductor diode lasers.

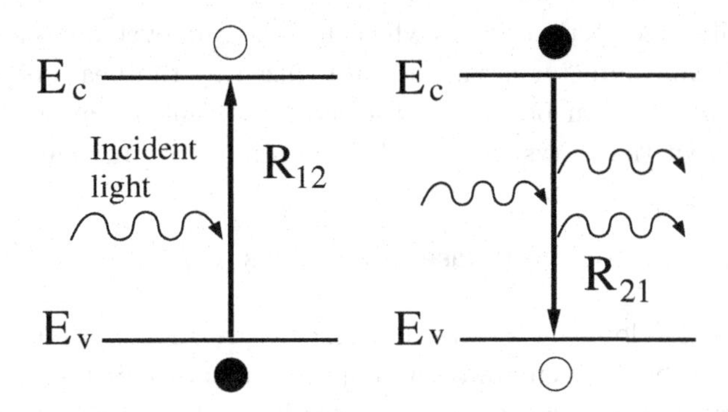

Fig. 2.5 Band-to-band stimulated absorption and emission, black dots represent electron, while grey ones holes.

For the radiation to be amplified, the net stimulated emission rate must be positive, that requires occupancy probability in the conduction band be larger than that in the valence band, viz., $f_c > f_v$, this means the population inversion in the semiconductor diode laser (necessary condition), and is the equivalent, in a semiconductor laser, to the conventional inversion condition $N_2 > N_1$. From the population inversion condition $f_c > f_v$, one can have $\Delta E_F = E_{Fc} - E_{Fv} > E_2 - E_1 = h\nu$, where ν is the frequency of emission photon. Therefore, only photons with frequencies whose photon energies hν are smaller than the quasi-Fermi level separation are amplified, and the net stimulated emission rate becomes positive. Since photon energy must at least be equal to the bandgap energy E_g, from this relation, we have

$$\Delta E_F = E_{Fc} - E_{Fv} > E_{21} = E_g + \frac{\hbar^2 k^2}{2m_e^*} + \frac{\hbar^2 k^2}{2m_h^*} > E_g, \tag{2.7}$$

where m_e^* and m_h^* are effective masses of an electron and a hole, respectively. This determines the critical condition for lasing, the gain is zero at the frequency where $E_{Fc} - E_{Fv} = h\nu$, at high frequencies, the semiconductor absorbs photon.

The values of E_{Fc} and E_{Fv} are influenced by pumping process, i.e., by the carrier density (N) of the electrons being excited to the conduction band. E_{Fc} increases and E_{Fv} decreases as N increases. N satisfying $E_{Fc}(N) - E_{Fv}(N) = E_g$ is named as the carrier density at transparency N_{tr}. Carriers are injected into the semiconductor material to make the free electron density at some threshold value N_{th} to be larger than N_{tr}, then semiconductor exhibits a net gain. When this active medium is placed in a suitable cavity, laser action occurs when this net gain overcomes loss. The pumping of semiconductor lasers can be realized by the beam of another laser, or by an electron beam, but the most convenient way is to apply electrical current that flows through the semiconductor junctions.

2.2 Semiconductor diode laser structures

Without the development of the heterostructure diode, semiconductor lasers would probably been always an impractical cryogenic laboratory devices. In this section, we first introduce bulk diode lasers including the homojunction, then heterojunction diode laser structures, and then investigate the one-dimension quantum well and zero-dimension quantum dots laser structures.

2.2.1 *Homojunction lasers*

Consider the homojunction diode laser, the pumping process is achieved across the p-n junction, where both p-type and n-type materials are in the form of semiconductor, as shown in the Fig. 2.6. Assume no current is applied to p-n junction, the band structure is schematically shown in the Fig. 2.6(a).

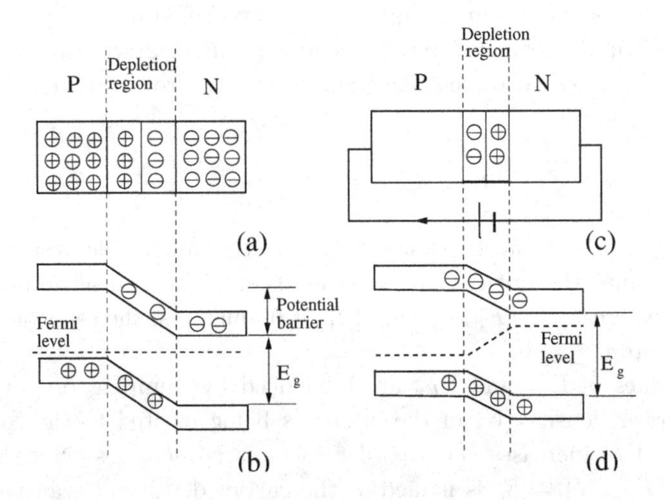

Fig. 2.6 Homojunction lasers, (a) open-circuited device; (b) energy vs. position diagram for open-circuited homojunction; (c) forward biasing; (d) energy versus position for forward biasing.

When two material is brought into contact, the Fermi level is not aligned, but in a very short time (~ps), majority carrier holes in the p region diffuse across the junction from p side to n side, similarly, majority carrier electrons in the n region diffuse across the junction from n side to p side. However, as these holes and electrons diffuse across the junction, many of them combine instantly, which leaves the p region populated with a few less holes, and similar for electrons in the n side. Therefore, it follows that a carrier-depletion region will exist on both sides of the junction, with the n side of the region positively charged and p side negatively charged, this is shown in Fig. 2.6(b). The charges on both sides of the depletion region generate an electric field across the region, the resultant electrical field opposes the diffusion of holes into n region and electron into the p region. By this way an energy barrier is produced and Fermi level is lined up, the height of energy barrier is called barrier voltage.

The dynamic equilibrium could be destroyed when a forward bias current is applied to the device as shown in Fig. 2.6(c). The two Fermi levels are misaligned (Fig. 2.6(d)). The injection of electrons into conduction band while holes into valence band reduce the barrier voltage, more of carriers are now able to cross the narrowed depletion region, many of electron-hole pairs combine radiatively and generate spontaneous emission. Therefore, for appropriate values of current density, the transparency and the laser threshold conditions can be satisfied. However, it is found that, for lasing, the densities of the electrons and holes in the conduction and valence band, respectively, are required to be significantly greater than $10^{18} cm^{-3}$. A major obstacle to achieving these densities in the p-n junctions formed with a single material is that carriers rapidly diffuse away from the junction, so a homojunction laser has a very high threshold current density at room temperature, which prevents practical use of this kind of semiconductor laser.

2.2.2 *Double-heterostructure lasers*

We have seen in the proceeding section that p-n junction can be lasing. Unfortunately, the current required to achieve lasing is of tens of thousands of Amperes per square centimeter, making it impractical for many applications. A more efficient solution is to use heterostructure and double heterostructures. A heterostructure is a junction formed by two different types of semiconductor, with one a larger bandgap than the other. When two semiconductor are put together, potential barriers are formed which can confine the electrons and holes, while the single heterojunction lasers do perform better than homojunction lasers, it is much more advantageous to extend heterojunction to double heterostructures. The better confinement of carriers and optical field resulted in the first successful continuously working diode laser at room temperature [Alferov *et. al.* (1970); Hayashi *et. al.* (1970); Thompson and Kirby (1973)].

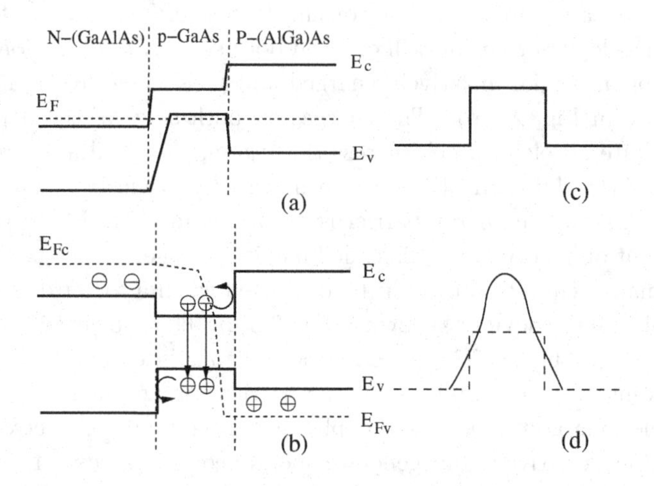

Fig. 2.7 Operation principle of double-heterojunction laser diode, (a) cross section structure, (b) effect of carrier confinement, (c) refractive index profile, (d) light intensity profile.

Figure 2.7 shows a n-p-p double heterostructure, when no current is applied, the Fermi-level is constant across the junction as show in Fig. 2.7(a). When the double heterostructure has been forward biased, the depletion region is reduced and the bands of the n-type shifts towards. The electrons are injected into the small bandgap active region from the n-doped cladding layer with higher bandgap E$_g$, while the holes enter the active region from the p-doped cladding layer on the opposite side. Because the active layer are sandwiched between the higher bandgap cladding layer, electrons cannot climb over the p-type cladding layer and holes cannot penetrate into the n-type layer. So the injected electrons and holes are confined in the active layer as illustrated in Fig. 2.7(b). Where population is inverted and it can combine radiatively and lase [carrier confinement as shown Fig. 2.7(c)]. The refractive index of active layer is higher by some percent that of cladding layer which confines the generated light with the p active layer (photon confinement in Fig. 2.7(d)). Light penetrating into the p and n cladding layer is not absorbed due to its wide bandgap. These three properties greatly decrease the threshold current density.

Because charge carriers in heterojunction lasers are confined to a much smaller region than in homojunction lasers, the change in index of refraction at the interface between the p GaAs and the p AlGaAs can provide a guiding effect for the laser beam. This is known as index guiding. Double heterojunction structure provides even more control over the size of the active region, and also provides additional index of refraction variation that allows for guiding of the optical wave.

2.2.3 *Quantum well lasers*

The carrier-confining effect of the double heterostructure is one of the most important features of modern diode lasers. With the advent of modern epitaxial growth techniques that allow one to fabricate active regions in the heterojunction structure on the atomic scales. It is possible to produce thin layers in semiconductor heterostructures, the so-called quantum well (QW) structures [Dingle *et. al.* (1974); Dupuis *et. al.* (1979); Zory (1993)] as shown in Fig. 2.8(a). Since the quantum effects occur only in one dimension, the energy level can be attained by considering an electron as a particle in one dimension potential well, it can be expressed by

$$E_n = \frac{n^2 \pi^2 \hbar^2}{2m^* w^2}, \tag{2.8}$$

where n= 1,2... is the energy state, m^* is the effective mass of a particle, and w is the length of the well. Hence, when the confined electrons in the conduction band and holes in the valence band combine, the photon is radiatively emitted with its frequency ν, which is given by

$$h\nu = E_g + \frac{n^2 \pi^2 \hbar^2}{2w^2}(1/m_e^* + 1/m_h^*), \tag{2.9}$$

where m_e^* and m_h^* are the effective masses of an electron and a hole, respectively.

Quantum well diode lasers and bulk double heterostructure lasers share many common characteristics, but with a thin core layer of the order of a few nanometers, in which quantization effect occurs and the density of states is a function of energy as shown in Fig. 2.8(b). Several advantages of quantum well lasers, compared with bulk lasers, can be deduced from Fig. 2.8. Fewer states have to be filled to reach population inversion. Therefore, the threshold current should be reduced. Further consideration shows that if only one energy level is filled, the spectral width of spontaneous emission and gain should be much smaller than for bulk lasers. QW laser diodes have gained many important applications for their good performance as compared to the conventional bulk laser diodes.

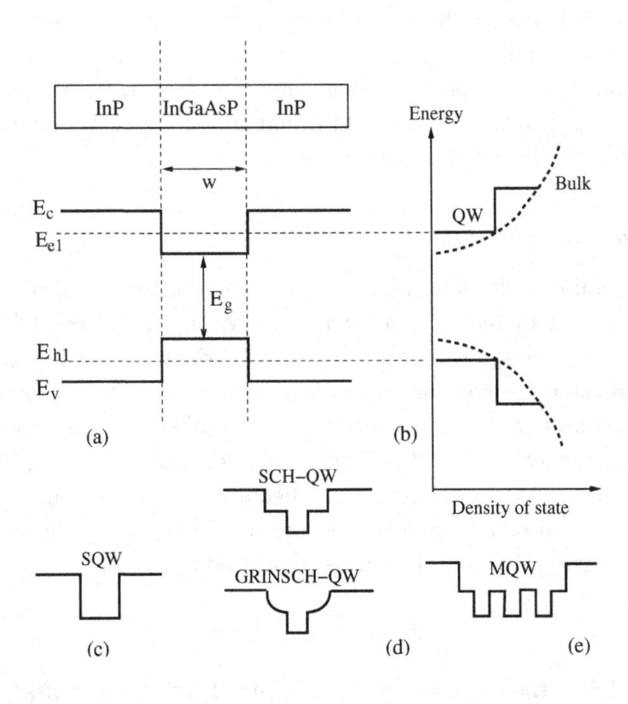

Fig. 2.8 (a)InGaAsP/InP single quantum well. Layer structure, energy band diagram. (b)Density of states of quantum well and bulk semiconductor (dashed lines). (c) Bandgap energies in a thin single QW, (d) in a SCH QW structure, and (e) in a MQW structure.

Now consider the operation of single quantum well (SQW) laser [Fig. 2.8(c)], if one would simply miniaturize the dimensions of a bulk double heterostructure to QW sizes, it would be pretty hard for carriers to be injected into the well, this would result in the increase of the threshold current and loss of optical gain. A novel approach to circumvent this is to separate the well from the cladding region by a composition in such a way that the barrier height is much less than the cladding's. This is called separate confinement heterostructure (SCH) as indicated Fig. 2.8(d). Alternative method is to modify the bandgap of the cladding region by the collection of carriers, this is called graded index separate confinement heterostructure (GRIN-SCH) laser. In this way, the optimization optical confinement and carrier confinement are effectively decoupled so that optical confinement can be significantly improved for typical QW lasers. Effective optical confinement and larger optical gain could also be achieved by multiple quantum well (MQW) structure as shown in Fig. 2.8(e).

2.2.4 *Quantum dots lasers*

Quantum dots, the ultimate in confinement, are still the subject of intensive research particularly in university laboratories in North America, Japan, and Europe. Quantum dots have been called artificial atoms [Harrison (1999)], despite the fact that they generally consist of hundreds of thousands of atoms. Confined in a dot or box, electrons should occupy discrete energy levels of the bottom of the conduction band and the top of the valence band. Lasing in devices confined in this way is therefore restricted to a much narrower range of wavelengths than in a conventional semiconductor, approaching the atomic ideal of infinitely narrow linewidths. The emission wavelength is determined by the dot size, which means that by controlling the size distribution, the range of emission wavelengths can be tailored for individual lasers. Restricting the effective bandgap also enhances the material gain and reduces the influence of temperature on laser performance. A QD laser therefore enables the practical applications of atomic physics in semiconductor devices. This leads to low operating currents due to an enhanced gain, very narrow linewidth, and minimal adverse effects with increased temperature [Reithmaier and Forchel (2002)].

Figure 2.9 illustrates four typical semiconductor laser crystal geometries and their corresponding functions for density of states versus energy. The quantum wires are also included. Table 2.1 summarizes the expressions of density of states for various semiconductor laser crystal structures, where m^* is effective mass of an electron, E is the total energy of an electron.

Table 2.1 Density of states for various dimension of semiconductor lasers.

Dimension	g(E)
3D(bulk)	$\propto \sqrt{E - E_0}$
2D(quantum well)	constant
1D(quantum wire)	$\propto \frac{1}{\sqrt{E-E_n}}$
0D(quantum dot)	$\propto \delta(E - E_n)$

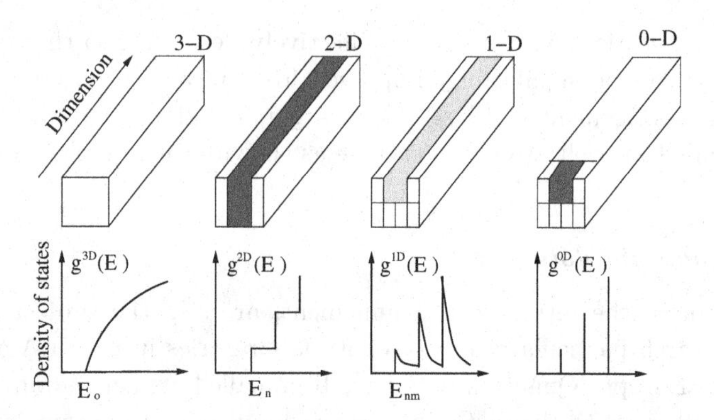

Fig. 2.9 Schematic of crystal geometries and expressions for density of states as a function of energy.

2.3 Basic characteristics of diode lasers

In this section we consider the threshold condition of Fabry-Perot diode laser and some elementary properties of diode laser characteristic such as output power, beam divergence, and spectral contents.

2.3.1 *Threshold condition*

Threshold current is the most important and basic parameter for laser diodes. Fig. 2.10 shows the structure of a general-purpose Fabry-Perot diode laser, which is modelling as a resonator containing plane optical waves travelling back and forth along the length of diode laser. Incident spontaneous emission light propagating to the reflection mirror is amplified by stimulated emission and comes back to initial position after a round trip inside the laser cavity. This process is subject to losses arising from light going through or diffracting at the reflection mirrors, and scattering or absorption within an active light-emitting medium. When the total loss is higher than the gain, the light attenuates. Injected current strengthens the amplification light in the diode laser, and when the gain and loss are balanced, initial light intensity becomes equal to the returned light intensity, this condition is referred to as threshold. A diode laser oscillates above the threshold when the gain is high enough.

Population inversion is not guaranteed to lase, in order to make the stimulated emission much larger than absorption, one has to build up a resonant cavity in which the light is reflected back and forth many times

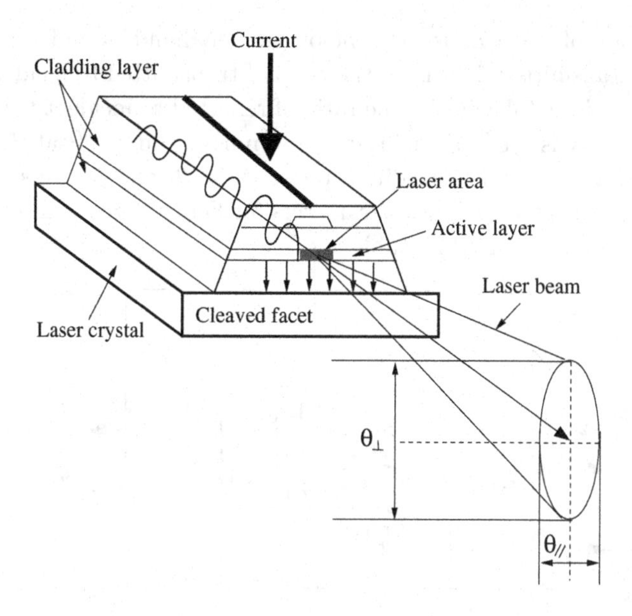

Fig. 2.10 Structure and lasing mode of laser diode.

before leaving the cavity. If the gain equals loss, lasing might occur in the resonant cavity.

Table 2.2 Typical parameters for semiconductor diode laser.

Parameter	Symbol	Typical value
wavelength	λ	0.4~ 2.0 μ m
small signal gain	g	$10^4 \sim 10^5$
cavity length	d	200~ 500μ m
index of refraction	n	3.4
confinement factor	Γ	0.2~ 0.5
mirror reflectivity	r_1, r_2	0.55
absorption coefficient	α	$45cm^{-1}$
operating temperature	T	300K
output power	P	milliwatt to Watt

Consider a semiconductor laser cavity of length d with a plane wave of complex propagation constant $k = \beta + j(g - \alpha)$, as shown in Fig. 2.11. α is internal attenuation per unit length, β is propagation coefficient, g is gain per unit length, that is produced by the stimulated effect, n is refractive index of cavity material, and λ is the wavelength of wave in the free-space.

The amplitude of wave E_i is incident on the left hand side of the cavity, the ratio of transmitted light is t_1, the ratio of transmitted to incident light fields at the right is taken as t_2, the ratio of reflected to incident fields with the optical cavity is $r_1e^{i\theta_1}$ at the right hand mirror and $r_2e^{i\theta_2}$ at the right-hand mirror. For a low loss medium, phase shifts θ_1 and θ_2 are small and are generally neglected. Typical parameters for Fabry-Perot semiconductor diode laser are summarized in Table 2.2.

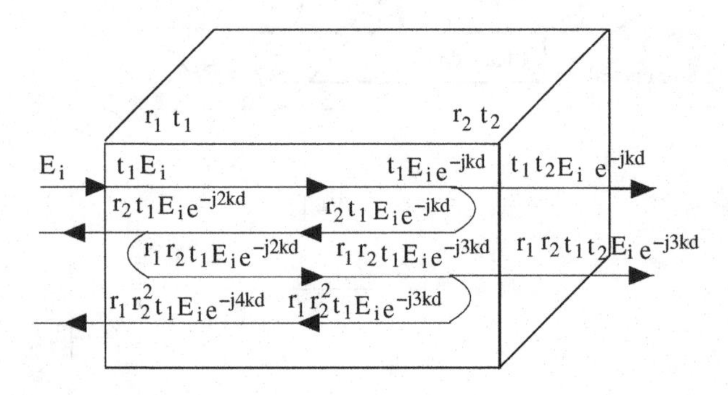

Fig. 2.11 Schematic of optical field propagation inside a diode laser cavity.

The plane wave electric field independent of time is expressed as E_ie^{-jkz}. At the first transition boundary, the light field E_i enters the cavity, transmits the first mirror, the field just inside left boundary becomes t_1E_i. The electrical field then proceeds into the laser cavity, in which the wave is reflected successively from one mirror to the other. The laser first arrives at the right boundary with filed $t_1E_ie^{-jkd}$. The first portion of the field transmitted output at the right boundary is $t_1t_2E_ie^{-jkd}$, and a portion $t_1r_2E_ie^{-jkd}$ is reflected back into the cavity toward to the left boundary, the next portion of the light transmitted output becomes $t_1t_2r_1r_2E_ie^{-j3kd}$ and so on. Adding up all the successive contributions, the output E_o is given:

$$E_o = t_1t_2e^{-jkd}(1 + r_1r_2e^{-j2kd} + (r_1r_2)^2e^{-j4kd} + ...), \tag{2.10}$$

the sum is a geometric progression which permits the last equation to be written as

$$E_o = E_i[\frac{t_1t_2e^{-jkd}}{1 - r_1r_2e^{-j2kd}}]. \tag{2.11}$$

When the denominator of Eq. (2.11) tends to zero, the condition of a finite transmitted wave E_o with zero E_i is obtained, which is the conditions for oscillation. Therefore the oscillation condition is reached when

$$r_1 r_2 e^{-j2kd} = 1. \tag{2.12}$$

Substituting the term k defined above in the gain medium, the resonance condition turns out to be

$$r_1 r_2 e^{2(g-\alpha)d} e^{-j2d\beta} = 1. \tag{2.13}$$

The condition for oscillation represents a wave making a round trip 2d inside the cavity to the starting plane with the same amplitude and phase, within an integer multiple of 2π, the general amplitude condition is:

$$r_1 r_2 e^{2(\Gamma g-\alpha)d} = 1, \tag{2.14}$$

where we have incorporated the confinement factor Γ inside the cavity into the gain, Eq. (2.14) is generally written as

$$\Gamma g_{th} = \alpha + \frac{1}{2d} \ln \frac{1}{r_1 r_2} = \alpha + \alpha_m. \tag{2.15}$$

This is the condition for threshold, where α_m is mirror loss. The gain coefficient per unit length strongly depends on the emission energy, on the operation conditions, and on light intensity in an active layer. If one replaces the r_1 and r_2 with power reflectivities R_1 and R_2, respectively, Eq. (2.15) can be rewritten by

$$\Gamma g_{th} = \alpha + \frac{1}{2d} \ln \frac{1}{\sqrt{R_1 R_2}}. \tag{2.16}$$

The phase condition is found to be:

$$\phi = 2d\beta = 2q\pi, \tag{2.17}$$

which reduces to

$$q = 2nd/\lambda, \tag{2.18}$$

where q is an integer, a resonance occurs when an integer number of half-wavelength λ fits into the cavity.

2.3.2 *Output power*

Output power is one of the important parameters to characterize a diode laser. Fig. 2.12 shows an experimental result, which depicts output power of a typical continuous wave (cw) in a semiconductor diode laser as a function of injection current (L-I curve).

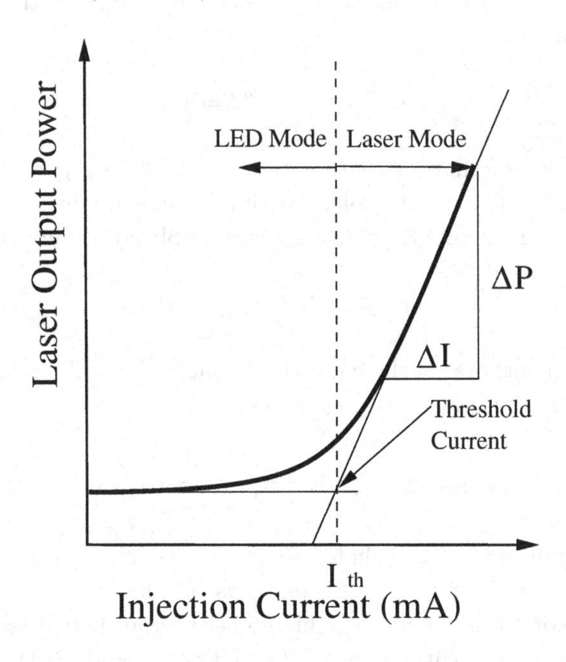

Fig. 2.12 Output light vs. current (L-I curve) characteristics for a diode laser.

When the forward bias current is low, the laser diode operates like a light-emitting diodes (LEDs) where the carrier density in the active layer is not high enough for population inversion, spontaneous emission is dominated in this region. As the bias increases, population inversion occurs, stimulated emission becomes dominant at a certain bias current, the current at this point is called threshold current. The injection current above the threshold induces the abrupt onset of laser action and coherent light is emitted from the diode laser. The laser threshold current is evaluated by extrapolating the linear part of the characteristic to zero output power. The slope of the increase in the lasing region is proportional to the differential external quantum efficiency as described below.

Lasing occurs for every wavelength mode as we can see from the

Eq. (2.18), the longitudinal mode separation $\Delta q = 1$ is given by differentiating Eq. 2.18), we have

$$\Delta\lambda_q = -\frac{\lambda^2}{2n_{eff}d} \tag{2.19}$$

where n_{eff} is an effective index refraction, which can be written as $n_{eff} = n[1 - (\lambda/n)(dn/d\lambda)]$.

In order to obtain the output power, the simple way to do is the rate equation from a single pair of coupled equations. Rate equations for carrier density N and photon density N_p of the i-th longitudinal mode oscillation read

$$\frac{dN}{dt} = \frac{I}{eV} - \frac{N}{\tau} - S(N - N_{th})N_p, \tag{2.20}$$

$$\frac{dN_p}{dt} = S(N - N_{th})N_p - \frac{N_p}{\tau_p}, \tag{2.21}$$

where τ is the recombination lifetime of carrier, τ_p is the photon lifetime in the cavity, V is the volume of gain medium, I is injection current, N_{th} is the threshold density of carrier, e is the electron charge, S is combination factors accounting for confinement factor, gain efficient, cavity volume, an other parameters. Under steady state condition. $dN/dt = dN_p/dt = 0$, we have the carrier density and photon density

$$N = N_{th} + \frac{1}{S\tau_p}, \tag{2.22}$$

and

$$N_p = \frac{\tau_p}{eV}(I - I_{th}), \tag{2.23}$$

where $I_{th} = \frac{eV}{\tau}N_{th}$ is threshold injection current, one finds the photon density above the threshold is a linear function of injection current I, the total photon energy E in the laser cavity within an active volume V is written as $E = N_pVh\nu$, where $h\nu$ is a photon energy. Therefore, the optical power emitted by stimulated emission in the cavity is

$$P_{in} = \eta_{in}\frac{h\nu}{e}(I - I_{th}), \tag{2.24}$$

where η_{in} is defined as the internal quantum efficiency, which is the fraction of carriers that radiatively recombined in the layer.

The practical interesting power is the output power emitted from each end mirror of the laser cavity, which reads

$$P_{out} = \eta_{in}\frac{h\nu}{e}(I - I_{th})\frac{\alpha_m}{\alpha_{tot}}, \tag{2.25}$$

where $\alpha_{tot} = \alpha + \alpha_m$ is the total loss of the laser. We can also define the external quantum efficiency η_{ex} as

$$\eta_{ex} = \frac{d(P/h\nu)}{d(I/e)} = \eta_{in}\frac{\alpha_m}{\alpha_{tot}}. \tag{2.26}$$

In addition to the external quantum efficiency, the slope efficiency of diode laser is also used quite often and given by

$$\eta_{slope} = \frac{dP}{dI} = \eta_{in}\frac{h\nu}{e}\frac{\alpha_m}{\alpha_{tot}}. \tag{2.27}$$

Now, if we use the slope efficiency, the output power of laser is simplified

$$P_{out} = \eta_{slope}(I - I_{th}). \tag{2.28}$$

If the injection current is increased to excessively high values, L-I curve becomes sub-linear. Operation at overly high currents shortens the lifetime and can fatally damage the laser.

2.3.3 *Beam divergence and astigmatism*

Divergence of output laser beam from a diode laser is described in Fig. 2.10. The beam is diffraction-limited in the plane of orthogonal and parallel of the junction due to the small size of the laser diode chip. Assume $d_\perp$ and $d_\parallel$ be the beam dimension (full width between 1/e points of the electric field) in the two directions, and suppose a Gaussian field distribution in both transverse directions. The beam divergence $\theta_\parallel$ in the plane parallel to the junction and $\theta_\perp$ in the plane orthogonal to the junction are give by $\theta_\parallel = 2\lambda/\pi d_\parallel$ and $\theta_\perp = 2\lambda/\pi d_\perp$ respectively, where λ is the lasing wavelength. For an output beam with an elliptical cross section (e.g., $\sim 1\mu m \times 5\mu m$), thus $\theta_\perp$ is larger than $\theta_\parallel$.

In many applications, for example, optical recording, near field of the laser is focused to an almost diffraction-limited spot. Many types of laser diodes have some astigmatism in the emitted beam. This is due to the presence of an optical loss at both sides of the laser stripe combined with the fact that the injected charge carriers in the stripe region contribute negatively to the refractive index. The magnitude of the astigmatic distance

D depends on the wavelength λ, the radius of curvature of the wavefront R, and the width of the near-field at the laser mirror w. The relation between these quantities for a Gaussian beam propagation can be given by

$$D = R/[1 + (\frac{\lambda R}{\pi w^2})^2]. \tag{2.29}$$

For gain-guided lasers, the dimensionless quantity $\lambda R/\pi w^2$ is comparable to or smaller than unity; for index-guided lasers, it is much larger than unity. R is usually of the order of $20 \sim 80\ \mu m$. It can be inferred from Eq.(2.29) that for gain-guided lasers, D≈R, whereas for index-guided lasers, D≪R. The question of which value of D are tolerable depends strongly on the numerical aperture used, and hence on the type of applications.

2.3.4 *Spectral contents*

The basic spectral characteristics of semiconductor diode laser with free running (solitary laser) have been studied in the last sections. As a last section in this chapter, we are going to examine the spectral contents of diode laser. Fig. 2.13 illustrates a schematic of experimental setup for examining the spectrum of a free-running diode laser.

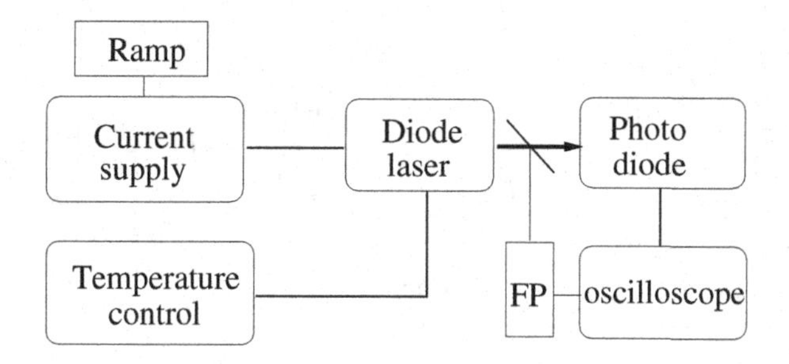

Fig. 2.13 Schematic of experiment setup for observing the spectrum of a free-running diode laser.

The diode laser (Sharp LTO24 MD specified output power 30 mW) is index-guided, single-mode laser, its temperature can be controlled by high precision temperature controller. Injection current is supplied by high precision current source, the current can be readily scanned by an external ramp voltage. Therefore, the output wavelength of diode laser can be tuned by changing the temperature and/or current. Fig. 2.14(a) shows the chip oscillation modes as a function of wavelength for a typical Sharp LTO24

780 nm laser diode. This spectrum was obtained with no external cavity and the laser running at a current just below threshold. All wavelengths that fit in the cavity by an integral number of half-wavelengths gain intensity. The crystal itself emits wavelengths over a large range but only those that fit properly will add constructively after reflection at the crystal's end facets. Fig. 2.14(b) shows the intensity of spectrum as Fabry-Perot interferometer is scanned, the free spectral range of this F-P cavity is 1.5 GHz. One can see that the line width of this laser diode is about 150 MHz.

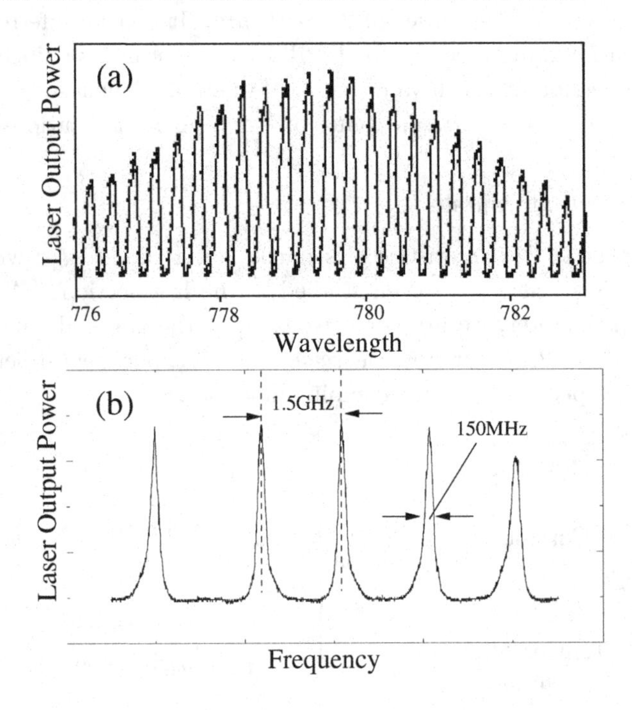

Fig. 2.14 Spectra of free-running uncoated diode laser, (a) intensity of spectrum as a function of wavelength, (b) intensity vs spectrum analysis of the output of laser obtained by scanning Fabry-Perot interferometer.

Chapter 3

Tunable Monolithic Semiconductor Diode Lasers

In this chapter, we review the state-of-the-art of monolithic single-mode and wavelength-tunable diode lasers, which have found many important applications in a variety of fields such as wavelength division multiplexing (WDM) technology in optical communication systems, advanced optical sensors, and optical test and measurement systems [Laude (1993)].

3.1 Introduction

Tunable semiconductor lasers continue to be of great interest for long time in scientific and engineering applications. Various designs are reviewed in this chapter with particular emphasis on the broadly tunable types in the application of coherent optical telecommunications [Coldren (2003)].

3.1.1 *DBR and DFB lasers*

We have introduced the Fabry-Perot (F-P) cavity diode lasers in the previous chapter, in order to make single-mode tunable lasers, one has to modify the two-facet Fabry-Perot laser structure. The cleaved-coupled-cavity or C^3 laser is a derivative of the F-P laser, where a F-P laser is cleaved into two shorter F-P laser of incommensurate lengths with the cleaver forming a third shorter (typically $< 1\mu m$) air-filled cavity between the two laser sections [Tsang *et. al.* (1983); Coldren and Koch (1984); Tsang (1985)]. Such lasers have many attractive properties such as single-mode operation with tunability. However, their manufacturability and stability in operation have been proved problematic in a real system. The distributed Bragg reflector (DBR) lasers and distributed feedback (DFB) lasers can fulfil this objective, this is one of approaches to achieving the tunable lasers by monolithic configuration.

A distributed feedback laser (DFB) consists of an active medium in which a periodic thickness variation is produced in one of the cladding layers that form the part of the heterostructure [Carroll *et. al.* (1998)]. The periodic spatial distribution of the index of refraction within the gain medium can be written as

$$n(z) = n_o + n_1 \sin[\frac{2q\pi z}{\Lambda} + \phi], \tag{3.1}$$

where q is any integer value, usually taking q=1, Λ is the pitch of the periodic thickness change, n_0 is the effective refractive index of the structure without the grating and much larger than n_1. The forward and backward propagation beams of the DFB laser are effectively coupled to each other if the Bragg condition is satisfied for the first order structure such that

$$\lambda = \lambda_B = 2n_0\Lambda, \tag{3.2}$$

wavelength can be selected by changing the refractive index and/or the pitch Λ in the medium. Fig. 3.1(a) shows a schematic of a DFB laser.

DBR lasers have the advantage, as compared to DFB lasers, that the grating is fabricated in an area separated from the active layer as shown in Fig. 3.1(b). Consider a dielectric waveguide as illustrated by Fig. 3.2(a). There is a corrugation in the dielectric region, which forms a grating of pitch Λ length per cycle, whose strength depends on the contrast of the indices and on the lithographed physical depth. The fundamental characteristics of optical-wave coupling with the grating can be described by the coupled mode equations, where there are two solutions of the optical wave equation, the two waves propagating in opposite directions. One wave is represented by ε_1, which is travelling in the +z direction with propagation constant $\beta_a = 2\pi n_a/\lambda$. The counter propagating wave is expressed by ε_2 with propagation constant $\beta_b = 2\pi n_b/\lambda$.

The resonance condition required by Bragg condition is

$$\Lambda = \lambda/2n_0, \tag{3.3}$$

it is required by the phase matching condition in the waveguide. The coupling of ε_1 and ε_2 can be described by the coupled mode equations [Yariv (1991)],

$$\frac{\partial \varepsilon_1}{\partial z} = \kappa^* \varepsilon_2 e^{-j2\Delta z}, \tag{3.4}$$

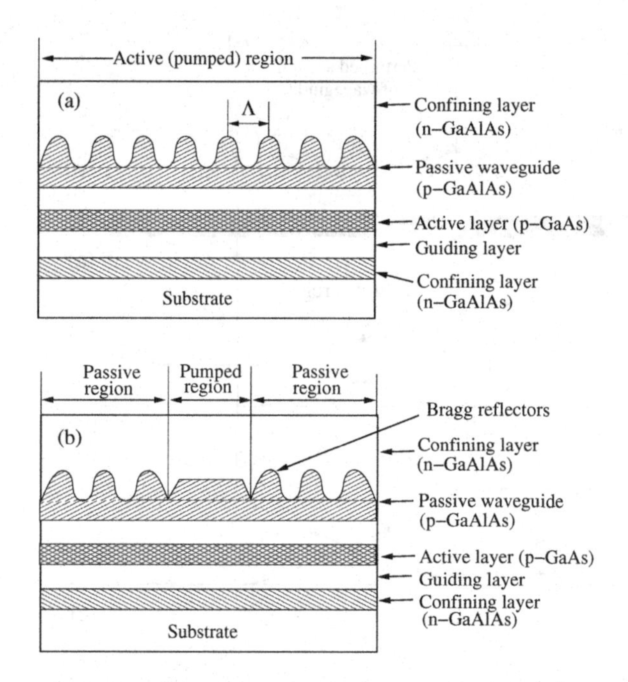

Fig. 3.1 Two typical laser structures using built-in frequency selective resonator gratings: (a) distributed feedback (DFB) laser, (b) distributed-Bragg-reflector (DBR) laser.

$$\frac{\partial \varepsilon_2}{\partial z} = \kappa \varepsilon_1 e^{j2\Delta z}, \tag{3.5}$$

where κ is coupling coefficient between the two modes, the parameter $\Delta = \beta_b - \beta_a - qk - j\gamma$ is the phase-mismatching, q is an integer, and γ is the exponential gain constant of the medium. The coupling of ε_1 into ε_2 is shown in Fig. 3.2(b) for the case where there is no gain in the region of the grating, and in Fig. 3.2(c) where there is. These two solutions of the coupled mode equations with different initial conditions correspond to the DBR ($\gamma = 0$) and DFB lasers as shown in Fig. 3.1.

In a DBR laser, the active region provides the gain, and the grating functions as wavelength selectivity, they are separated along z. In a distributed feedback (DFB) laser, the two functions are combined, where the grating is actually along the entire length l of the active region, two waves are getting amplified in both directions of z. Fig. 3.2(b) and (c) show the change of the intensity of two wave as a function of z in both directions.

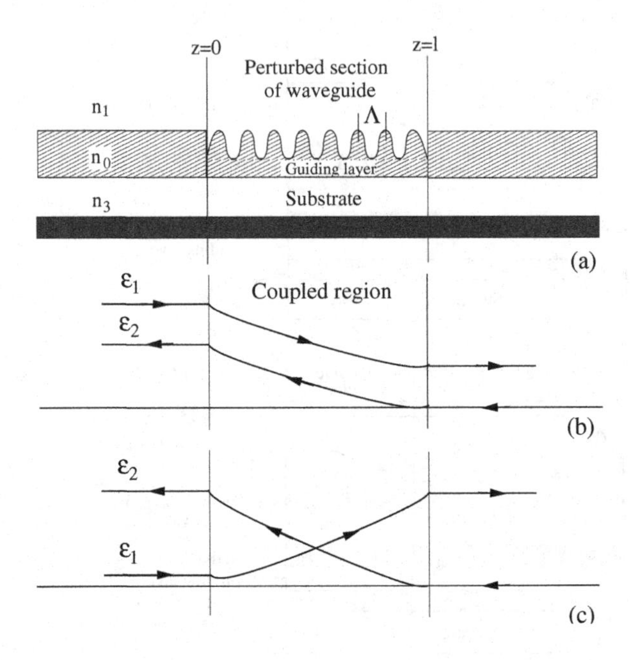

Fig. 3.2 Grating structures for single-mode lasers. (a) Grating is a slab waveguide. (b) Field strengths of forward and counter propagating wave, no gain. (c) Same as (b) with gain.

However, in a perfect DFB laser, there are actually two modes produced, a way of avoiding this is to have both ends AR coated, the added reflection acts to make device asymmetric and suppress one of the spectral lines. In this way, the DFB will be used as a resonant filter so that there are many passes back and forth through the grating. In order to have maximum transmission exactly at Bragg resonance, the grating must be modified by adding a $\lambda/4$ phase shift in the middle section of the grating as shown in Fig. 3.3(a). This shift introduces a sharp transmission fringe into the grating reflection band in Fig. 3.3(b) (right), which narrows the linewidth of the laser significantly.

3.2 Tunable monolithic diode lasers

In this section, we introduce the various methods to achieve continuous tunable single-mode wavelength in monolithic DBR and DFB lasers [Westbrook *et. al.* (1984); Dutta *et. al.* (1986); Dutta *et. al.* (1986);

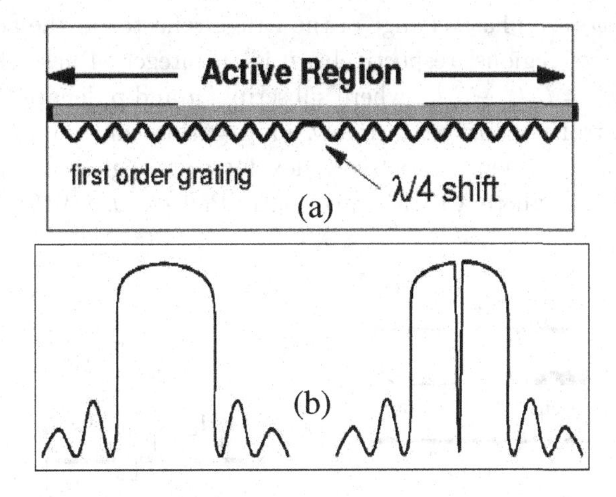

Fig. 3.3 Grating with quarter wavelength ($\lambda/4$). (a) Quarter wavelength shifted gating, (b) reflection characteristic spectrum (left) without phase shift (right) with phase shift.

Murata *et. al.* (1987); Coldren and Corzine (1987)]. Tunable monolithic semiconductor lasers are developed based on the DBR and DFB diode lasers as described in the proceeding section, where a waveguide corrugation, which forms a grating, functions as a basic element of single mode selection and tunability [Kobayashi and Mito (1988)], DBR diode laser incorporates separate active and passive grating reflector region. There are three configurations to achieve the tunable diode lasers, which are DBR diode laser with Bragg wavelength control, those with phase control and those with Bragg wavelength and phase control, the combination of control of Bragg wavelength and phase results in a widely tunable range. In DFB diode laser, phase control for the light reflected from one facet or non-uniform excitation along the cavity causes a change in the lasing condition, which gives rise to the wavelength tunability.

3.2.1 *Distributed Bragg reflector diode laser*

The principle of how to tune the DBR diode laser was discussed by Kobayashi and Mito [Kobayashi and Mito (1988)]. The schematic diagram is illustrated in Fig. 3.4. The lasing mode must satisfy the phase matching condition:

$$\phi_1 = \phi_2 + 2q\pi, \tag{3.6}$$

where ϕ_1 and ϕ_2 are phase change of the Bragg reflector in the active and the phase control regions, respectively. q is an integer. Phase change ϕ_2 is written as $\phi_2 = \beta_a l_a + \beta_p l_p$, where subscripts a and p denote the active and phase control regions, respectively, propagation constants β_a and β_p depend on the equivalent refractive index for each region n_a and n_p as $\beta_{a,p} = (2\pi/\lambda)n_{a,p}$, where λ is the wavelength [Pan *et. al.* (1988)].

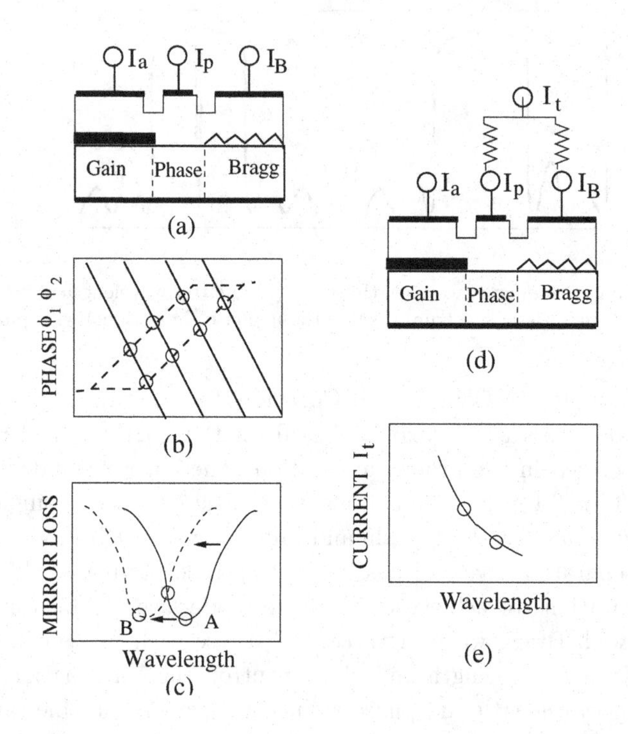

Fig. 3.4 A diagram for operating principles for a DBR laser diode based tunable LD. (a) Fundamental configuration. (b) Phase versus wavelength. (c) Mirror loss versus wavelength. (d) Connection of the Bragg reflector region and the phase control region to a current source through resistors. (e) Wavelength vs total current. Adapted with permission from *J. Lightwave Tech.* **6**, 11, pp. 1623-1633. Kobayashi and Mito (1988).

The phase change in the Bragg reflector can be derived from complex reflectivity r for the Bragg reflector [Yariv (1991)] as

$$r = \frac{-j\kappa \sinh \gamma l}{\gamma \cosh \gamma l + (\alpha_i + j\Delta\beta) \sinh \gamma l} = |r| \exp(j\phi_1), \tag{3.7}$$

here,

$$\gamma^2 = \kappa^2 + (\alpha_i + j\Delta\beta)^2, \tag{3.8}$$

$$\Delta\beta = \beta - \beta_0 = \frac{2\pi}{\lambda} n_b - \frac{\pi}{\Lambda}, \tag{3.9}$$

where $\Lambda, \kappa, \alpha_i$ and n_b denote the corrugation period, the corrugation coupling coefficient, the loss, and the equivalent refractive index for the Bragg reflector, respectively, l is the total cavity length. Since the Bragg reflectivity profile with wavelength is a function of $\beta = 2\pi n_b/\lambda$, a change of Δn_b in n_b yields a wavelength shift in the reflectivity profile peak, which is simply given by

$$\frac{\Delta\lambda}{\lambda} = \frac{\Delta n_b}{n_b}. \tag{3.10}$$

Thus, the Bragg wavelength can be changed by varying the refractive index of layers near the corrugation.

The possible lasing modes are given by the points where the ϕ_1 and $\phi_2 + 2q\pi$ curves cross. As shown in Fig. 3.4(b). The lasing mode will be the highest total gain with the lowest mirror loss mode, mirror loss with the Bragg reflector reflectivity can be written as

$$\alpha_m = \frac{1}{2l} ln(\frac{1}{rR}), \tag{3.11}$$

where R is the output facet reflectivity. Fig. 3.4(c) shows the mirror loss dependence on the lasing wavelength, these curves could be shifted with phase and mirror loss by varying the injection current in each region. Therefore the wavelength of DBR could be tuned. In a specific example as shown in Fig. 3.4(c), the lasing mode changes from mode A to mode B with mode jumping. Three operating schemes are considered in the following, depending on the combination of Bragg wavelength and phase control.

3.2.1.1 *Bragg wavelength control*

When only the Bragg wavelength is controlled, the lasing wavelength moves to the shorter wavelength with mode jumping. The process is demonstrated in Fig. 3.5(a). Assume any mode A on one of many phase lines $\phi_2 + 2q\pi$, it moves towards the shorter wavelength on this line, when it reaches the point A′ where ϕ_1 line and ϕ_2 line intersects. Another mode B on another ϕ_2 line has same mirror loss, beyond this point, mode B has lower mirror

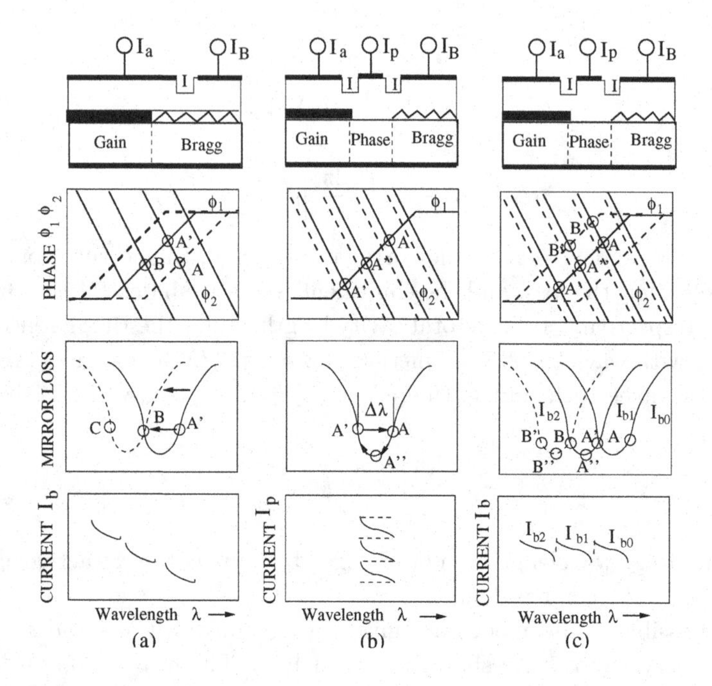

Fig. 3.5 A diagram for a tunable DBR with (a) Bragg wavelength control, (b) phase control, (c) Bragg wavelength and phase control. I: Isolator. Adapted with permission from *J. Lightwave Tech.* **6**, 11, pp. 1623-1633. Kobayashi and Mito (1988).

loss than mode A, and begins to lase, which results in the mode hopping. The continuous tuning range can be enlarged by reducing the active region length with respective to the total cavity length, because the slope of phase ϕ_2 line is proportional to the active region length.

3.2.1.2 *Phase control*

It is possible to shift the phase ϕ_2 line in parallel toward the shorter wavelength by means of the injection current into a phase control region. As show in Fig. 3.5(b), here the mirror loss curve and phase ϕ_1 curve for the Bragg reflector are fixed. The mode A has the same mirror loss as mode A$'$, with the current increasing, the lasing mode moves in the cyclic way as $A \rightarrow A'' \rightarrow A' \rightarrow A$. Thus, the wavelength tunable range is limited to the value $\Delta\lambda$, as depicted in Fig. 3.5(b).

3.2.1.3 *Bragg wavelength and phase control*

A wider tuning range will be achieved by combining the two schemes of Bragg wavelength control and phase control as described above. The fundamental configuration is shown in Fig. 3.5(c), in this way, two various operating schemes are involved, i): quasi-continuous tuning; ii): continuous tuning.

i) Quasi-continuous tuning: this scheme is defined as continuous wavelength tuning with different longitudinal modes. The tuning method is shown in Fig. 3.5(c). When mode A comes to the point A′, in similar way to the phase control, the Bragg wavelength is changed in such a way that the new mode B coincides in wavelength with the previous A′. By varying the phase again, mode B begins to lase and moves to the B′ through mode B″. By applying the phase control current in a repeated fashion and the Bragg wavelength control current in the step-like fashion, the lasing wavelength can be varied over a wide range, as shown in Fig. 3.5(c).

The maximum tuning range in the quasi-continuous mode in the three-section DBR laser diode is mainly limited by the Bragg wavelength variation range, which is determined by the change in the refractive index for the Bragg reflector. Another factor limiting the tuning range is the threshold current increase and the light output power decrease caused by the loss increase due to free carrier absorption and by the heating in cw operation with the injected carrier density increase in Bragg reflector region and the phase control region. This kind of limitation is intrinsic because when a carrier injection scheme is applied, the refractive index changes through the plasma effect [Okuda and Onaka (1977)].

ii) Continuous tuning: some changes in both the Bragg wavelength and the phase simultaneously makes it possible to tune lasing wavelength continuously, while keeping the identical longitudinal cavity mode, this is shown in Fig. 3.4(d) and (e). If the change of Bragg wavelength $\Delta\lambda_b$ is equal to that in phase $\Delta\lambda_p$, the emission wavelength can be varied without changing the other lasing condition. Suppose that the refractive index change with current is proportional to the square root of the injection current, injection current to the Bragg reflector and the phase control regions should meet the requirement by

$$\frac{I_b}{I_p} = (\frac{l_a}{l_a + l_b})^2 (\frac{l_a}{l_b}), \tag{3.12}$$

this relation can be satisfied by applying the injection current from a single current source through a suitable dividing resistor to each electrode as shown in Fig. 3.4(d).

More recently, Debregeas-Silard [Debregeas-Silard *et. al.* (2002)] have demonstrated for the first time an integrated DBR modular with two-section and simple operating conditions, that can provide a constant 20 mW coupled output power and SMSR above 40 dB over 40× 50 GHz spaced channel.

3.2.2 *Distributed feedback diode laser*

The operation principle of tuning DFB laser diode is not as simple as DBR's, because it is difficult to control the phase and the Bragg wavelength separately in DFB diode laser. However, the basic idea for tuning is identical, i.e., to change the lasing conditions to satisfy phase matching condition. Two schemes have been proposed to achieve the lasing conditions, (a): corrugation phase control, and (b): nonuniform excitation along the cavity.

3.2.2.1 *Corrugation phase control*

The lasing condition for DFB lasers depends on the corrugation phase at the facet where the reflectivity is not small enough. If the corrugation phase is varied, the lasing mode moves on a locus in the threshold gain versus propagation constant. The corrugation phase can be equivalently controlled by changing the phase of light reflected from the facet [Kitamura *et. al.* (1985); Murata *et. al.* (1987)]. With the variation of the corrugation phase at the facet, lasing mode moves towards the shorter wavelength until another mode at the longer wavelength reaches the same threshold gain as the original mode. Beyond this point, the lasing mode jumps to the longer wavelength mode, which leads to mode hopping and repeats the cyclic wavelength change with the tuning current.

3.2.2.2 *Multielectrode*

Nonuniform excitation along the cavity in DFB laser enables a change in the lasing condition, Multi-electrode DFB lasers have been realized to achieve nonuniform excitation[Kuznetsov (1988)], where the electrode is divided into two or three sections along the cavity. The tuning function is provided by varying the injection current ratio into the two sections. Typical geometries are shown in Fig. 3.1(b), tuning in these devices results from the combined effects of shifting the effective Bragg wavelength of the grating in one or several sections, together with the accompanying change of the optical path length from the change of refractive index itself.

DFB and DBR lasers based on Indium phosphide (InP) have been matured due to the wide applications in 1300 and 1550 nm telecommunication wave band. To attain shorter wavelength of 770 to 860 nm DFB and DBR lasers which are useful in exciting and cooling cesium and rubidium atoms for atomic clock, laser cooling and BEC. Gallium arsenide (GaAs) technology combined with aluminium (Al), In, and P alloying must utilized. Wenzel et al [Wenzel *et. al.* (2004)] presented a ridge-waveguide GaAsP/AlGaAs laser, emitting an optical power of up to 200 mW in a single lateral and longitudinal mode at a wavelength of 783 nm. The distributed feedback is provided by a second-order grating, formed into an InGaP/GaAsP/InGaP multilayer structure. The laser is well suited as a light source for Raman spectroscopy and even as primary sources in frequency conversion systems.

3.2.3 *Summary of tunable monolithic diode laser*

To summarize the monolithic laser diode developed in the last sections, typical examples are listed in Table 3.1 with configurations, operating principles and tuning ranges.

Table 3.1 Typical examples of monolithic laser diodes.

Configuration	Operating principle	Tuning range	References
Bragg wavelength control	$\Delta\lambda_B$	4 nm	[Woodward *et. al.* (1992)]
Phase control	$\Delta\phi$	5.8 nm	[Tohmori *et. al.* (1993)]
Bragg wavelength and phase control	$\Delta\lambda_B + \Delta\phi$	2.4 nm	[Murata *et. al.* (1987)]

3.3 Widely tunable diode lasers

It can be seen in the proceeding sections that the tuning range $\Delta\lambda$ is limited by the relationship $\Delta\lambda/\lambda = \Delta n/n$, where Δn is the change of refractive index and n is the index of the tuning section. In general, the tuning range is limited within 7 nm at 1.55 μ m, although the use of thermal effect [Woodward *et. al.* (1992)] has extended this to 22 nm [Oberg *et. al.* (1991)]. The tuning range could be broadened because the gain curve in the semiconductor can be as wide as 100 nm. In this section, we consider several ways to extend the tuning range beyond the $\Delta\lambda/\lambda = \Delta n/n$, one can have the tuning range $\Delta\lambda/\lambda = \Delta n/n - \Delta q/q + \Delta l/l$, where Δl is the change of cavity length, and Δq is varied by mode-selection filter. A generic tunable single-frequency laser is illustrated schematically in Fig. 3.6 [Coldren (2003)]. In this way, one might achieve the full range tuning of semiconductor lasers. A number of novel widely tun-

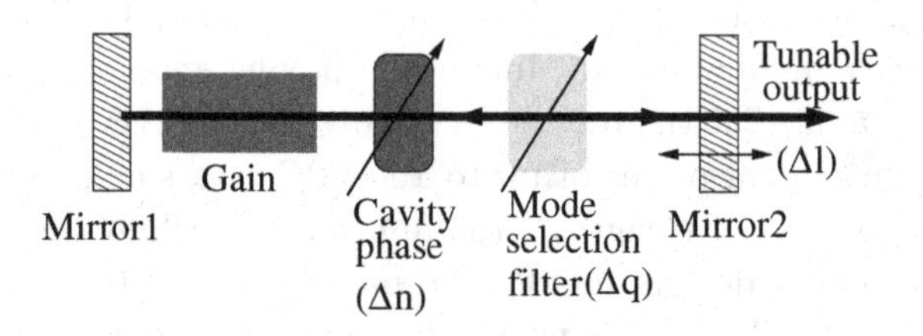

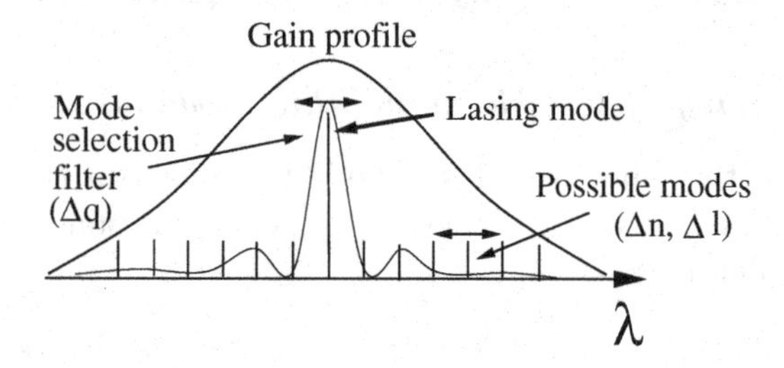

Fig. 3.6 Schematic of generic tunable single-frequency laser.

able semiconductor lasers have been developed in the early 1990s [Coldren (2000)]. These include the configurations that employ two multielement mirrors and Vernier-effect tuning enhancement [Jayaraman *et. al.* (1992); Tohmori *et. al.* (1993)] structure using grating-assisted codirectional couplers for enhancement tuning [Kim *et. al.* (1994)], some Y-branch configurations[Lang *et. al.* (1987); Hildebrand *et. al.* (1993)], and recently, VCSEL with movable mirrors [Larson and Harris (1996); Li *et. al.* (1998); Chang-Hasnain (2000)].

3.3.1 *DBR-type lasers*

It is possible to achieve an extended tuning range by use of a structure with multiple cavities and of a short grating as reflector for each cavity. Sampled grating and super-structure grating DBR lasers turn this possibility into reality.

3.3.1.1 *Sampled grating DBR (SGDBR) lasers*

Jayaraman et al. [Jayaraman *et. al.* (1993)] gave the first detailed description of widely tunable DBR lasers using sampled-grating theoretically and experimentally, they achieved the tuning range of 57 nm limited by the beat period. Other groups have also reported wide tuning range of 72 nm using a SG-DBR lasers [Mason *et. al.* (1997); Mason *et. al.* (2000)]. Fig. 3.7 shows the sampled-grating DBR (SGDBR) laser, which was the first monolithic device to ever tune over a wavelength range of 30 nm [Jayaraman *et. al.* (1992)]. The originally proposed four

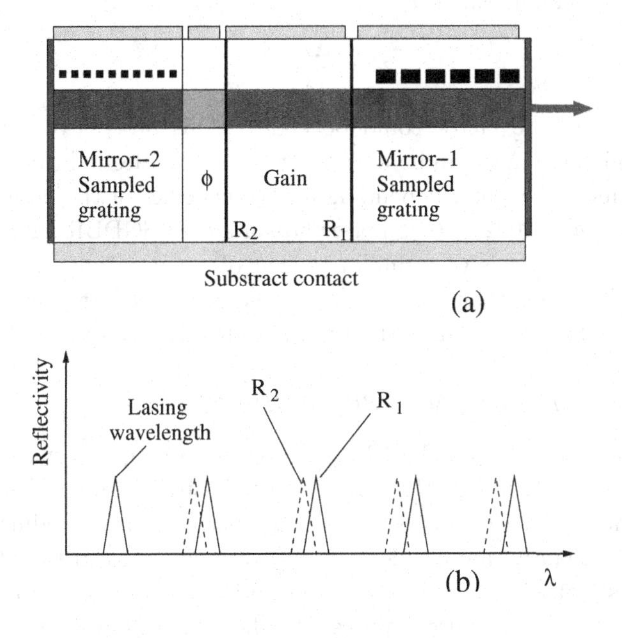

Fig. 3.7 Tunable laser based on mirrors with periodic spectra. (a) Schematic of sampled-grating DBR, (b) Vernier effect, two reflection combs from mirror 1 and 2.

section design and Vernier mirror tuning concept, along with the early discontinuous tuning concept that used only three-section, are illustrated in Fig. 3.7. The sampled-grating design uses two different multielement mirrors to create two reflection combs with the different wavelength spacings.

The laser operates at a wavelength which a reflection peak from each mirror coincides. Since the peak spacings are different for various mirror, only one pair of peaks can line up at a time, the reflection peaks are separated by

$$\Delta\lambda = \frac{\lambda^2}{2n_{eff}\Lambda}, \tag{3.13}$$

where n_{eff} is the effective index of structure, and Λ is the length of grating sampling period. This is typically chosen to be a little less than the available direct index tuning range, i.e., 6~8 nm so that the range between peaks can be accessed by tuning both mirrors together. Since the difference in peak spacing for either mirror is $\delta\lambda$, only a small differential tuning is required to line up the adjacent reflection peaks. Thus by Vernier effect, one can have the tuning enhancement factor by

$$F = \frac{\Delta\lambda}{\delta\lambda}, \tag{3.14}$$

for typical value, the factor could be eight. This process of equal and differential mirror tuning can be used to cover the full desired tuning range of typical value 40~80 nm, depending on the specifics of the design [Fish (2001); Shi *et. al.* (2002)]. One major advantage of SGDBR laser is that the fabrication process is very similar to that of a conventional DBR laser, the main difference is that for the same value of the peak reflectivity, the passive section has to be longer than the net effective cavity length.

3.3.1.2 *Superstructure-grating DBR (SSGDBR)lasers*

Most of the features of the SGDBR are shared by the super-structure grating DBR (SSGDBR) design. In this case, the desired multi-peaked reflection spectrum of each mirror is created by using a phase modulation of grating rather than an amplitude modulation function as in the SGDBR. Periodic bursts of a grating with chirped period are typically used. This multielement mirror structure requires a smaller grating depth and can provide an arbitrary mirror peak amplitude distribution if the grating chirping is controlled [Ishii *et. al.* (1996)]. However, the formation of this grating is very complicated. The SSGDBR was the first widely tunable laser structure to provide full wavelength coverage over more than 30 nm with good single mode suppression ratio [Ishii *et. al.* (1994)]. This range was improved to over 60 nm with further refinement [Ishii *et. al.* (1996)]. The structure of SSGDBR laser is shown in Fig. 3.8, The laser consists of four segments: a 600 μm-long active region, a 125 μm-long phase-control region, a 400 μm-long front SSGDBR region, and a 600 μm-long rear SSGDBR region. A 1.58 μm strained 6 quantum wells was used in the active region.

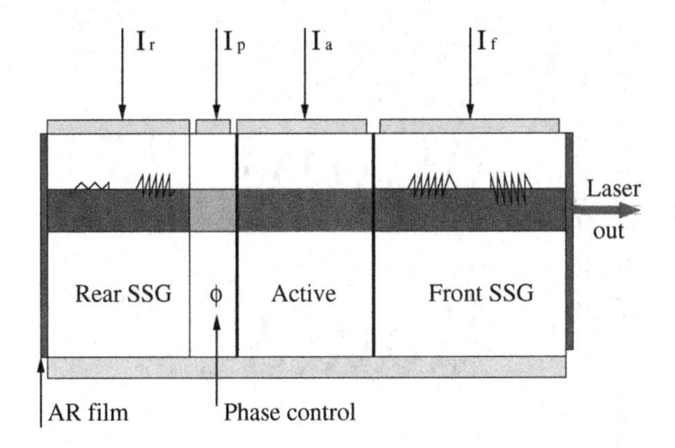

Fig. 3.8 Schematic structure of SSG DBR. Adapted with permission from *Electron. Lett.* **32**, 5, pp. 454. Ishii et al. (1996).

The phase of the diffraction grating in the DBR regions is periodically modulated to create multiple peaks in the reflection spectra. As the reflection peak spacing of the front SSG is different from that of the rear one, the total reflection spectrum has a maximum of the wavelength at which the peaks are lined up with each other, and hence single mode operation is achieved. Wide tuning can be obtained based on the principle of Vernier effect as shown in Fig. 3.7(b). The tuning range of $\Delta\lambda_t$ is approximately given by $\Delta\lambda_t = \lambda_f\lambda_r/(\lambda_f - \lambda_r)$, where λ_f and λ_r are the peak spacings of the front and rear SSG, respectively, and assuming $\lambda_f > \lambda_r$. Experimental results are shown in Fig. 3.9, the tuning currents to three ports, the output power from the front facet, and the single mode suppression ratio (SMSR) as a function of lasing wavelength was studied. A quasi-continuous wavelength tuning range of 62.4 nm was obtained with maximum power variation 10 dB. However, output powers are limited by the excessive mirror losses.

3.3.2 *Grating-assisting co-directional coupler (GACC-DBR)*

Very soon after the first viable SGDBR was produced, the first 1550 nm lasers incorporating grating-assisted codirectional- couplers (GACC) were demonstrated [Alferness *et. al.* (1992)]. A discontinuous tuning range of 57 nm was reported. However, the single mode suppression ratio (SMSR) was <25 dB over much of the range. This was soon improved to more acceptable levels, but sporadic tuning characteristics of this vertical-coupler

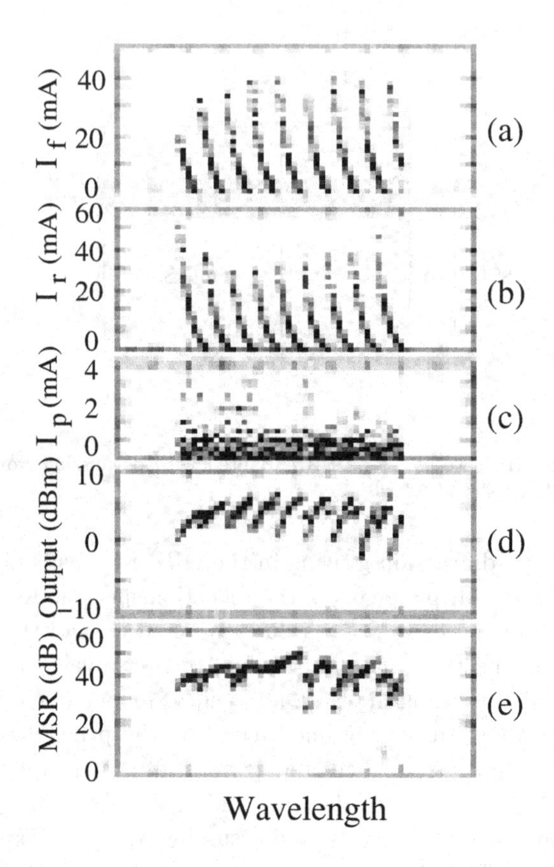

Fig. 3.9 Wavelength tuning characteristics of 1.55 μm SSG-DBR CW laser with a temperature of 30 $°C$, I_a=100 mA. Adapted with permission from *Electron. Lett.* **32**, 5, pp. 455. Ishii et al. (1996).

filter (VCF) laser seemed to be inherent. As shown in Fig. 3.10, the wide tuning range in this case is due to the enhanced tuning of the filter peak in a GACC, which is placed in the center of the VCF laser [Kim *et. al.* (1994)]. This enhanced tuning derives from the fact that the center frequency tunes as the index change in one guide relative to the difference in modal indexes between the two guides, rather than relative to the starting index, as in grating mirror or other phase shifting elements [Chuang and Coldren (1993)]. Thus, the tuning enhancement factor is given by $F = n_{1g}/(n_{1g} - n_{2g})$, the index of the tunable guide divided by the difference in modal group indexes. This factor can be easily ten times or more. Unfortunately, the GACC fil-

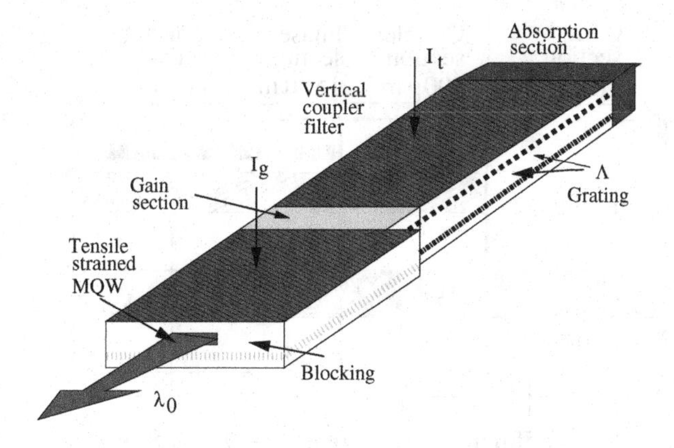

Fig. 3.10 Schematic of the broadly tunable VCF tensile strained MQW laser. Adapted with permission from *Appl. Phys. Lett.* **64**, 21, pp. 2764. Kim et al. (1994).

ter bandwidth also is proportional to F. Therefore, as F is increased, more axial modes of the laser tend to fall under the filter passband and become candidates for lasing.

The CW light-current characteristic of the laser with no timing current exhibits a threshold current of 30 mA with a $\sim$ 9% external differential quantum efficiency. Output light is TM- polarized because TM gain dominates over TE gain in the tensile-strained quantum well material. A tuning range of 74.4 nm with side mode suppression ratio of >25 dB and as high as 34 dB have been obtained by injecting as much as a 500 mA tuning current and applying a -3.4 V reverse-biased voltage into the VCF section. The entire wavelength tuning was achieved under CW operation.

3.3.3 *Grating-coupled sampled-reflector (GCSR)*

The solution to the GACC filter bandwidth problem is to add a SGDBR grating for one mirror in combination with the GACC. This approach was found to provide much better SMSR because near-in modes are not reflected by the SGDBR, the enhanced tuning of GACC does not translate into greater sensitivity to tuning current noise, etc.. The first good result with this so called GCSR (grating coupled sampled-reflector) laser was reported by Oberg et al. [Oberg *et. al.* (1991)]. A SSG-DBR back mirror was used in intermediate work to show full coverage over a range of 40 nm and 114 nm discontinuous tuning [Rigole *et. al.* (1995)].

As shown in Fig. 3.11, the GCSR structure is relatively complex with

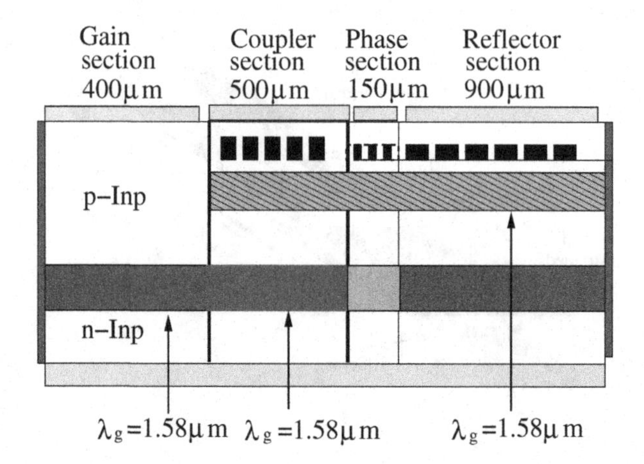

Fig. 3.11 Schematic of GCSR laser. Adapted with permission from *IEEE J. Photon. Tech. Lett.* **7**, 7, pp. 697. Rigole et al. (1995).

two vertical waveguide, three different bandgap regions, three changes in lateral structure, and different gratings. It is significantly more difficult to fabricate than the SGDBR laser. All wavelength measurement were made at a constant gain current of 250 mA and a fixed heat sink temperature of 20 °C. The tuning characteristic of the laser can be divided in three parts. Coarse tuning is obtained by changing the coupler current alone, the laser is tuned in large steps corresponding to the peak separation of the SSG-DBR, Fig. 3.12 shows the coarse tuning at zero bias of the SSG-DBR section (Bragg current). The total tuning range is more than 100 nm, wherein all the possible 20 peaks are accessed.

All the peaks of the SSG-DBR can be selected by a suitable current in the coupler with side mode suppression better than 20 dB, as it is shown in Fig. 3.13. The separation of the peaks of the comb varies between 5.7 nm and 4.6 nm, the variation in frequency spacing is mainly due to the dispersion of the group index. The output power at 250 mA with no tuning current reaches 4 mW and varies with 12 dB over the tuning range [Rigole *et. al.* (1995a)]. The medium tuning is performed when the coupler current and reflector current (Bragg current) are changed simultaneously, in this way, the selected peak is tuned at the same time as it is tracked for maximum coupler transmission by slightly increasing the current in the coupler. Apart from the coupler current change, the medium tuning is similar to the tuning of the Bragg wavelength in a DBR laser. By changing the current

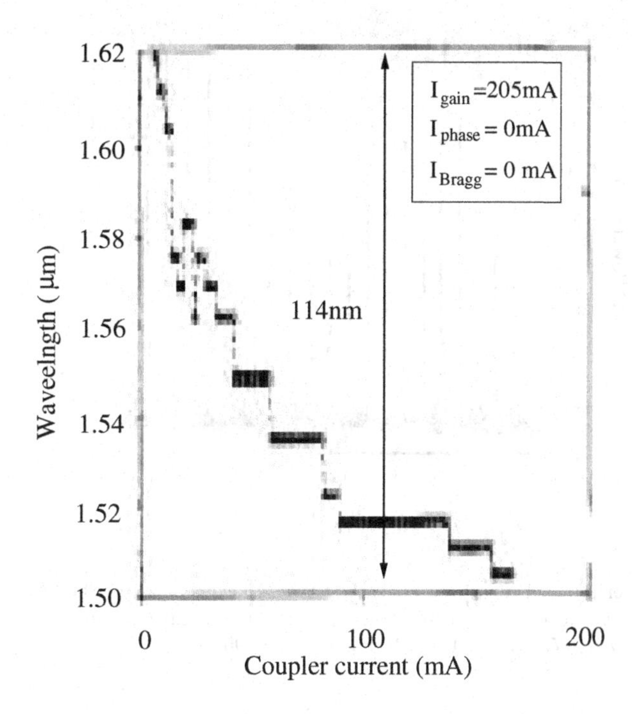

Fig. 3.12 Measured coarse tuning of the GCSR laser. Adapted with permission from *IEEE J. Photon. Tech. Lett.* **7**, 7, pp. 698. Rigole et al. (1995).

in the phase section, the lasing wavelength can be continuously fine tuned over the longitudinal mode spacing.

Combining the three types of tuning, one can access 16 wavelengths spaced by 1 nm from 1545 nm to 1560 nm as shown in Fig. 3.14. The current was set manually using conventional current source with 0.1 mA current resolution and monitoring the wavelength on an optical spectrum analyzer with a wavelength resolution of 0.1 nm, the SMSR was from 20 dB to 35 dB. It can be seen that a multi-section DBR laser typically requires for or more electrodes to achieve a wide tuning range and a full coverage of wavelengths in the range. The tuning characteristic is quasi- or discontinuous, and typically contains many steps. One advantages of a multi-section DBR laser structure is its easy integration with other devices such as a modulator, amplifier, and coupler.

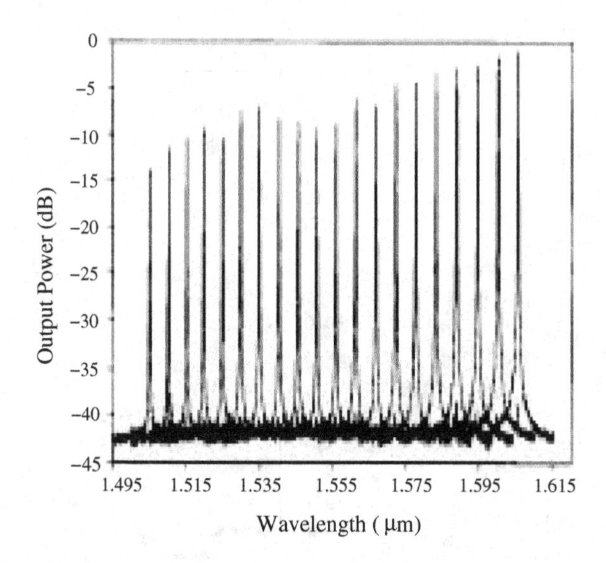

Fig. 3.13 Superimposed measured spectra of the 20 lasing peaks of the SSG-DBR obtained by tuning the coupler current only. Adapted with permission from *IEEE J. Photon. Tech. Lett.* **7**, 11, pp. 1250. Rigole et al. (1995).

3.3.4 *Diode laser arrays*

Laser arrays, where each laser in the array operates at a particular wavelength, are an alternate to tunable lasers, these arrays incorporate a combiner element, which makes it possible to couple the output to a single fiber. If each laser in the array can be tuned by an amount exceeding the wavelength difference between the array elements, a very wide total wavelength range can be achieved. The DFB laser diode array might be the most promising configuration for WDM optical communication systems due to its stable and highly reliable single-mode operation.

A wavelength laser array with an integrated amplifier and modulator designed for transmission of a single selectable wavelength has been demonstrated [Young *et. al.* (1995)]. The lasers have thresholds of $\sim$20 mA, and the single-step printed grating provided $\lambda/4$-shifted DFB lasers with an average channel spacing of 200 GHz. Output powers through the modulator of several milliwatts have been obtained from all six lasers when 75 mA is applied to the amplifier. A 12 nm tunable source with up to 15 mW fiber coupled output power has been fabricated by integrating four DFB lasers and a booster amplifier to provide a single output based on the technique

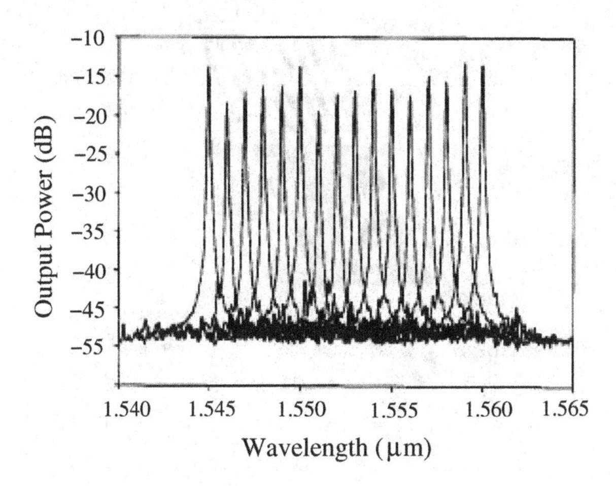

Fig. 3.14 Superimposed measured spectra of the 16 wavelengths spaced by 1 *nm*, obtained by controlling the three tuning currents. Adapted with permission from *IEEE J. Photon. Tech. Lett.* **7**, 11, pp. 1251. Rigole et al. (1995).

of high-yield and low-cost. The chip is tunable over 30 ITU channels with 50 GHz spacing [Pezeshki *et. al.* (2000)].

Recently, Kudo et al. [Kudo *et. al.* (2000)] have developed compact eight-channel wavelength-selectable microarray distributed feedback laser diodes with a monolithically integrated 8×1 multimode-interference (MMI) optical combiner, a semiconductor optical amplifier (SOA), and an electro-absorption (EA) modulator. Figure 3.15 schematically shows the scheme of the eight-microarray DFB laser diode wavelength-selectable microarray. The laser section is 400 μm long and 80 μm wide. The MMI section is 640 μm long and 80 μm wide. The SOA and MOD sections are 500 μm and 250 μm long, respectively. A $\lambda/4$-shift grating is introduced to the DFB cavity and a 25 μm long window structure with an antireflection coating was also made on the rear facet of the laser diode and the front of the modulator, respectively. In such a way, we can yield the high single longitudinal modes. Continuous L-I characteristics for the eight channel LD is shown in Fig. 3.16 with fixed SOA current 100 mA and no biased MOD current at a temperature of 25 $°C$. Uniform L-I characteristic with an average threshold current as low as 13.3 mA and an average output power of 6.9 mW at I_{DFB}=100 mA were obtained. Fig. 3.17 shows the corresponding eight-channel lasing spectra measured at I_{DFB}=50 mA. Stable

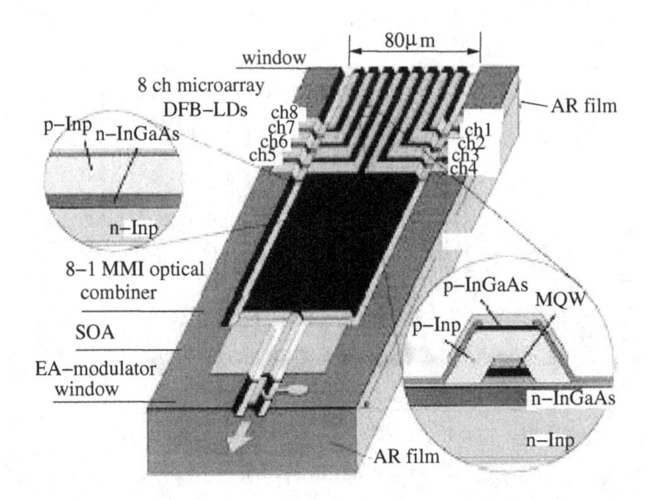

Fig. 3.15 Schematic of wavelength-selectable eight-stripe microarray DFB-LD's with monolithically MMI, SOA, and MOD. Adapted with permission from *IEEE Photon. Tech. Lett.* **12**, 3, pp. 242-244. Kudo et al. (2000).

single-longitudinal-mode operations with a SMSR of >40 dB and an optical signal-to-noise ratio (SNR) of 0.43 dB were obtained for all channels.

Figure 3.18 shows the wavelength tuning characteristics, a CW tunable wavelength range of 15.3 nm is obtained by switching the LD channels and changing the operating temperature from 5 °C to 25 °C. The tunable wavelength ranging from 1552.5 nm to 1567.8 nm covers 40 WDM channels with a 50 GHz spacing between each. The SMSR of >40 dB was maintained over the entire wavelength range.

A commercial laser diode arrays has been developed for simultaneous generation of laser output at two wavelengths at a grazing-incidence grating external cavity. By moving vertically with respect to the optical axis V-shaped double slit at the end mirror, tuning of the spectral separation of the dual-wavelength laser output from 3.52 to 11.29 nm has been demonstrated [Wang and Pan (1994)]. The side-mode suppression ratio is 10~20 dB.

3.3.5 *Vertical-cavity surface-emitting lasers (VCSEL)*

It is worthwhile to mention another monolithic tunable laser: tunable vertical cavity surface emitting lasers (VCSELs). The advantages of VCSEL include easier fiber coupling, simpler packaging and testing, and the ability to be fabricated in arrays. Vertical-cavity surface-emitting lasers (VCSEL)

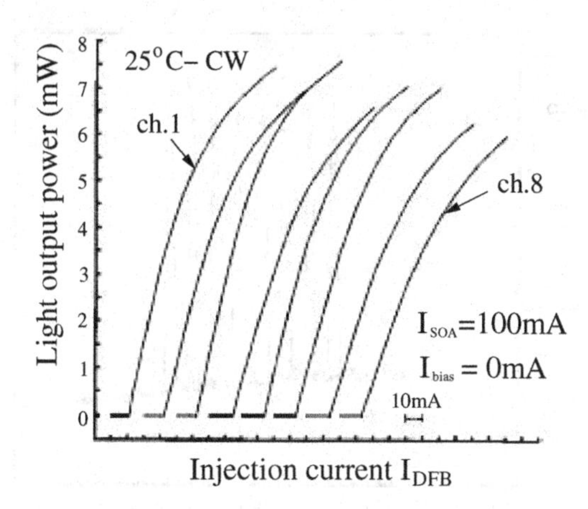

Fig. 3.16 Eight-channel L-I characteristics. Adapted with permission from *IEEE Photon. Tech. Lett.* **12**, 3, pp. 242-244. Kudo et al. (2000).

are now routinely used in photonic applications such as telecommunications, primarily at wavelength of 850 nm. Their long-wavelength devices, at wavelengths of 1300 nm and 1550 nm, still have much room to improve before they achieve such success. VCSEL can be classified roughly into three major categories based on mechanical and optical designs: cantilever VCSEL [Chang-Hasnain (2000)], membrane VCSEL [Yokouchi *et. al.* (1992); Larson *et. al.* (1995); Harris (2000)], and a half-symmetric cavity VCSEL [Vakhshoori *et. al.* (1997)]. In this section we mainly introduce the cantilever VCSEL, then briefly examine the deformable membrane VCSEL.

A tunable cantilever vertical cavity surface emitting laser (c-VCSEL) with a dozens of nanometer wavelength tuning for an applied voltage of several volts has been reported, both blue and red shift have been obtained [Wu *et. al.* (1995); Li *et. al.* (1997)]. Fig. 3.19 shows a schematic diagram of the tunable c-VCSEL. The top reflector of the laser cavity is formed as a composite mirror including a movable top DBR in a freely suspended cantilever, a variable air spacer layer and a fixed DBR. The bottom reflector of the laser cavity is a stationary fixed DBR. By applying a reverse biased tuning voltage between the top n-DBR and p-DBR across the air gap to create electrostatic attraction, the cantilever may be deflected towards the substrate, thereby changing the air spacer thickness and consequently

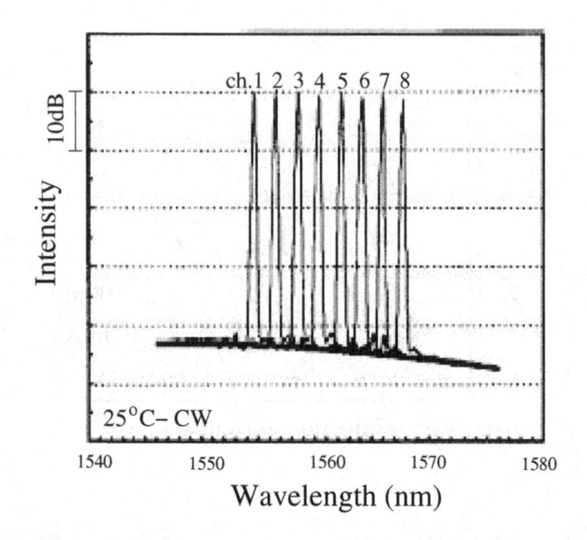

Fig. 3.17 Eight-channel lasing spectra from modulator facets at I_{DFB}=100 mA. Adapted with permission from *IEEE Photon. Tech. Lett.* **12**, 3, pp. 242-244. Kudo et al. (2000).

changing the resonant wavelength of the cavity.

The c-VCSEL lasing spectra is shown in Fig. 3.20 for different tuning voltage V_t when the active region is pumped at $1.2I_{th}$ [Li *et. al.* (1998)]. As the tuning voltage increases, the lasing wavelength blue-shifts and reaches a minimum of 919.1 *nm* at 19.8 V. The next order Fabry-Perot mode at 950.7 nm becomes the lasing mode at tuning voltage of 20.1 V. Increase of the tuning voltage blue-shifts the lasing mode until the tuning voltage reaches 26.1 V and lasing wavelength returns to the starting wavelength. A wide tuning range of 31.6 nm centered at 950 nm was achieved with the laser under room temperature of 22 °C continuous-wave operation. Fig. 3.21 shows the measured and calculated VCSEL wavelengths as a function of the tuning voltage from a tunable VCSEL with a single transverse mode, excellent agreement is achieved between the experimental measurements and theoretical calculations. The monotonic and well-behaved tuning curve confirm that a very simple wavelength locking mechanism can be sufficient to insure wavelength accuracy during tuning, that is a distinct advantage of a tunable c-VCSEL.

Recently fabricated MEMS VCSEL with a bridge structure has been shown its mechanical characteristics superior to those of cantilevered VC-

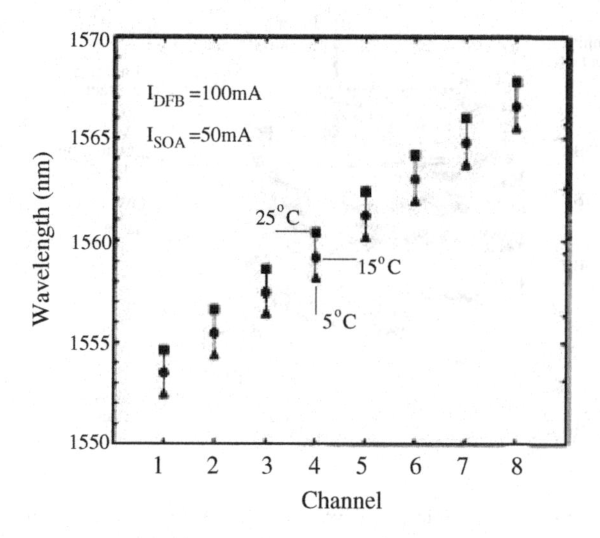

Fig. 3.18 Wavelength tuning characteristics obtained by laser diode channel selection and temperature control of ±10 °C. Adapted with permission from *IEEE Photon. Tech. Lett.* **12**, 3, pp. 242-244. Kudo et al. (2000).

SEL as described above. Both VCSELs have a top distributed Bragg reflector suspended above the body of the laser, their wavelength can be tuned by applying an electrostatic force between the suspended DBR and the body of laser. The VCSEL with bridge structure is designed to avoid the instability of the cantilevered VCSEL and to make it insensitive to mechanical vibration. The laser could be tuned over most of the C-band from 1543 nm to 1565 nm with a tuning voltage of 46 V, the wavelength is relatively linear with voltage, and the maximum output of power is approximately 1.3 mW [Sun *et. al.* (2004)].

An alternative approach for VCSEL similar to the previous one is to use a deformable membrane instead of cantilever, as shown in Fig. 3.22. Continuous wavelength tuning of 15 nm was achieved micro-electromechanically in a vertical cavity surface emitting laser operating near 960 nm with a micromachined deformable-membrane top mirror suspended by an air gap above a p-i-n diode quantum well active region and bottom mirror [Larson and Harris (1996); Larson *et. al.* (1998)]. Applied membrane-substrate bias produce an electrostatic force which reduces the air gap thickness, and therefore tunes the lasing wavelength.

It has been reported [Sugihwo *et. al.* (1998)] that an improved tunable

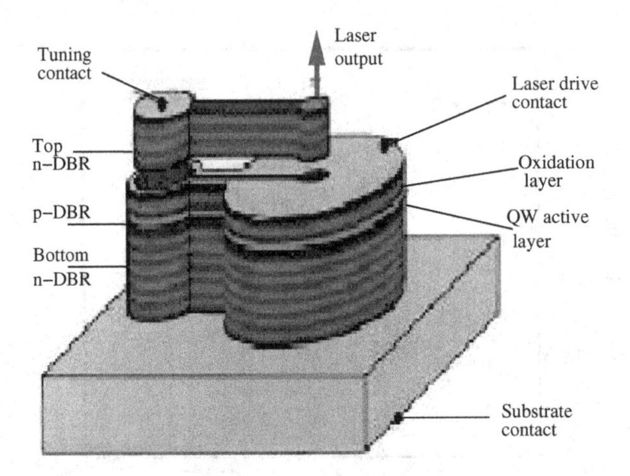

Fig. 3.19 Schematic of a tunable cantilever VCSEL. Adapted with permission from *IEEE J. Select. Topics on Quantum Electron.* **6**, 6, pp. 979. Chang-Hasnain (2000).

structure that incorporates a partial anti-reflection coating to increase coupling between the air gap and the semiconductor cavity. A more flexible micromachined process has been modified that enables independent optimization of the central reflector region and deformable membrane structure. This combination of structural and process modification makes it possible to decouple the tradeoffs between wavelength tuning rate and threshold current, as well as the tradeoffs between top mirror reflectance and tuning voltage. With these improved approaches, single-mode devices with a 30 nm wavelength tuning range have been produced by a 2.5 pair dielectric distributed Bragg reflector hybrid membrane top mirror.

Recently Kner [Kner *et. al* (2003)] presented the performance of a microelectromechanical system tunable vertical-cavity surface-emitting laser operating at 1550 nm and incorporating a tunnel junction for improved current injunction and reduced optical loss. These lasers show single-mode output power of 0.28 mW and a tuning range of 10 nm.

3.3.6 *Other widely tunable monolithic diode lasers*

There are many different types of widely tunable monolithic diode lasers, here we introduce the double-ring resonant coupled lasers recently reported lensless tunable external cavity lasers.

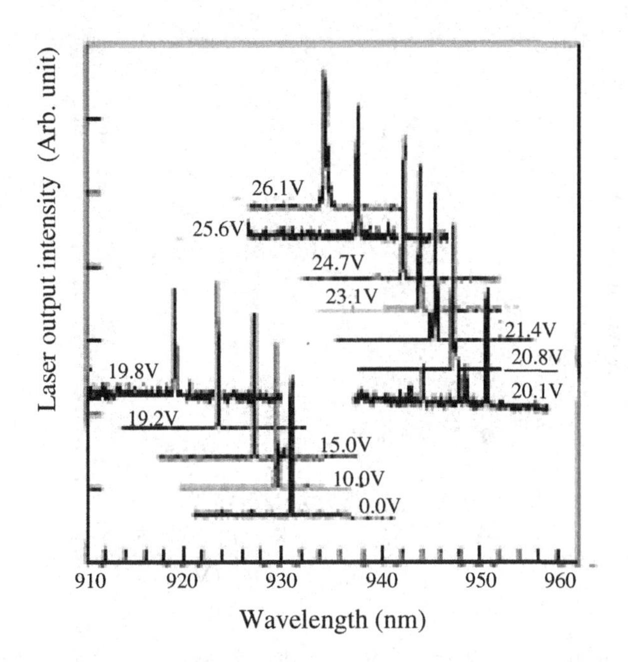

Fig. 3.20 The tuning spectra of a large aperture top-emitting c-VCSEL under various tuning bias. Adapted with permission from *IEEE Photo. Tech. Lett.* **10**, 1, pp. 19. Li et al. (2000).

3.3.6.1 *Double-ring resonant coupled lasers*

So far, we have introduced the monolithic structure lasers of sampled-grating distributed Bragg reflector lasers, super-structure grating DBR lasers, and VCSELs. A novel wavelength tunable source called double-ring resonant coupled lasers (DR-RCL) has been proposed and analyzed [Liu *et. al.* (2002)]. Benefiting from the uniform peak transmission, narrow bandwidth and other superior characteristics of travelling wave supported high-Q resonators, DR-RCLs offers many promising advantages over the conventional tunable lasers, including ultra wide wavelength tuning range, high side mode suppression ratio, uniform threshold and efficiency, narrow linewidth, low frequency chirp, and simple fabrication.

Figure 3.23 shows the proposed tunable double micro-ring resonator coupled laser structure, this laser is composed of four regions: gain region, two passive micro-ring resonators, passive waveguide, and absorption regions. The gain region provides light amplification. The two passive micro-ring resonators have slight different radii providing the mode selec-

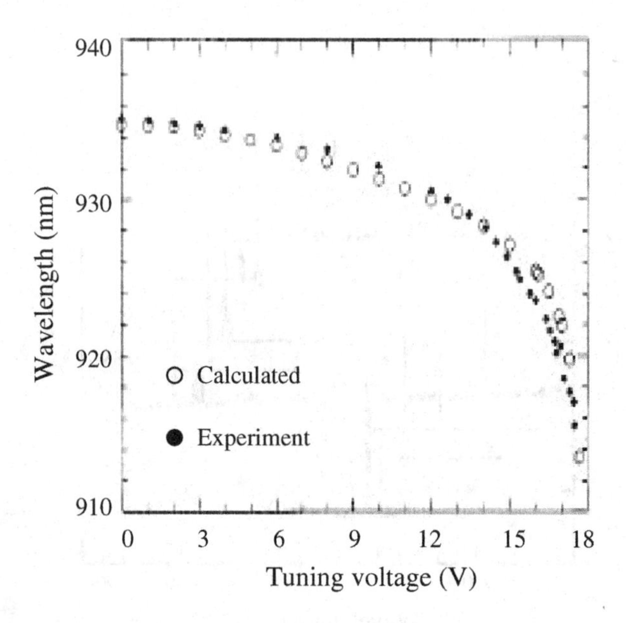

Fig. 3.21 Measured and calculated VCSEL wavelengths as a function of tuning voltage from a tunable VCSEL with a single transverse mode. Adapted with permission from *IEEE J. Select. Topics on Quantum Electron.* **6**, 6, pp. 982. Chang-Hasnain (2000).

tion and wavelength tuning mechanism. Four passive waveguide connect the ring resonators and the gain region. They could also serve as fine tuning phase regions. The absorption region extinguish the light possible back reflection from the facets and the rings. The front and back reflective facets form a laser cavity. The asymmetric double ring coupled laser is similar to the SG-DBR laser by replacing the front and back sampled-grating with two ring resonators. Compared to standing-wave supported SG-DBR structures, travelling-wave supported ring resonators have superior characteristics: very high-Q, narrow filter bandwidth and very large effective light travelling length.

The ring resonator provides a strong mode selection filtering. Only light at the resonance wavelength can be effectively coupled from gain region to the front or back passive waveguide and reflected by one or both facets to the gain region via the ring resonator again. Fig. 3.24 illustrates the basic tuning idea of double ring resonator coupled lasers. Each ring resonator has a set of transmission spectra, the wavelength period is the free spectra range FSR $=\lambda^2/(\mathrm{nl})$, where n is the effective index of the ring and l is the

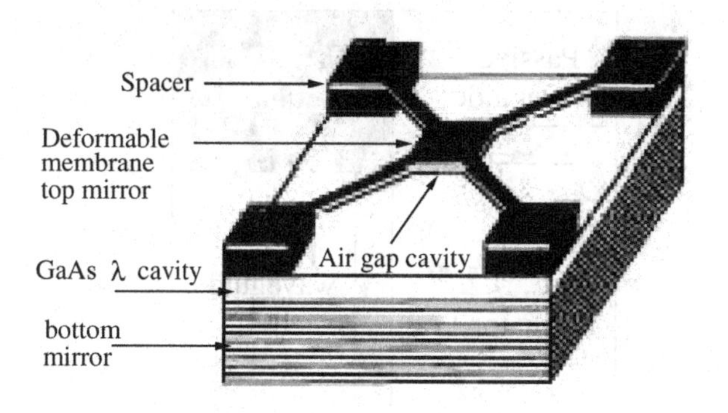

Fig. 3.22 Schematic of deformable membrane VCSEL. Adapted with permission from *IEEE Photon. Tech. Lett.* **7**, 4, pp. 382. Larson et al. (1995).

round trip distance in the ring. The two rings have slight different radii. Therefore, the two sets of transmission peak combs have small different peak spacing.

Similar to SG and SSG-DBR lasers, wavelength tuning is achieved by aligning the peaks in the two sets of combs with adjustment of index in one or both ring resonators in DR-RCLs. The wavelength tuning enhancement factor is expressed by $F=\Delta\lambda_g/FSR$, where $\Delta\lambda_g$ is the half bandwidth of the material gain. Compared to SG and SSG-DBR lasers, double ring resonant coupled lasers could offer a much larger tuning enhancement because of uniform peak transmission and ultra-narrow bandwidth.

3.3.6.2 *Lensless tunable external cavity lasers*

A novel lensless tunable external cavity laser using monolithically integrated tapered amplifier, grating coupler and external half mirror has been proposed and demonstrated [Uemukai *et. al.* (2000)]. It can be fabricated by a simple process and emits a collimated output beam. The schematic diagram of the propose tunable external cavity laser is shown in Fig. 3.25. The semiconductor device is constructed with a tapered amplifier and a grating coupler, it is fabricated from an InGaAs-AlGaAs single-quantum-well graded-index separate-confinement heterostructure waveguide. The laser cavity consists of a facet mirror, the tampered amplifier, the grating coupler, and the external half mirror. By rotating the semiconductor device with respect to the half mirror as shown in Fig. 3.25, the lasing wavelength can be tuned.

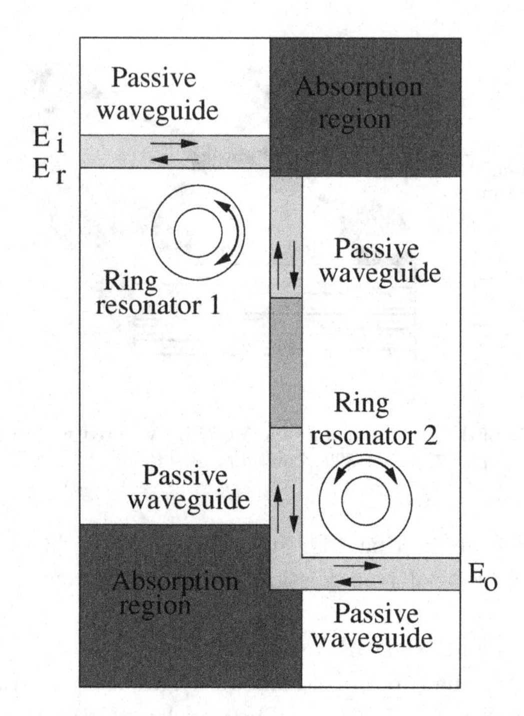

Fig. 3.23 Double ring resonator coupled laser structure. Adapted with permission from *IEEE Photo. Tech. Lett.* **14**, 5, pp. 600. Liu et al. (2002).

The device was mounted on a rotating stage, an half mirror with a 50% reflectivity was aligned at a distance of 10 mm, which forms the grating coupler. For all the measurements, the heat-sink temperature is fixed at 10 °C by a thermoelectric cooler. Fig. 3.26 shows the dependence of the output power through the half mirror on the injection current measured for CW lasing at exit angle of 11.5 °. The threshold current was about 1.0 A, and the maximum output power was 84 mW. The inset shows the lasing spectrum at I=1.8 A, the lasing wavelength was 1002.0 nm, and the linewidth was 0.08 nm. Wavelength tuning characteristics are examined by rotating the device as shown in Fig. 3.25. The dependence of the lasing wavelength and the output power on the device angle for I_{TA}=1.8 A and I_{NC}=3 mA is shown in Fig. 3.27. By varying the device angle from 13.0^o to 17.5^o, linear and continuous wavelength tuning over a wide range of 21.1 nm from 985.2 nm to 1006.3 nm was achieved.

In conclusion, we summarize the characteristics of monolithic tunable diode lasers in Table 3.2.

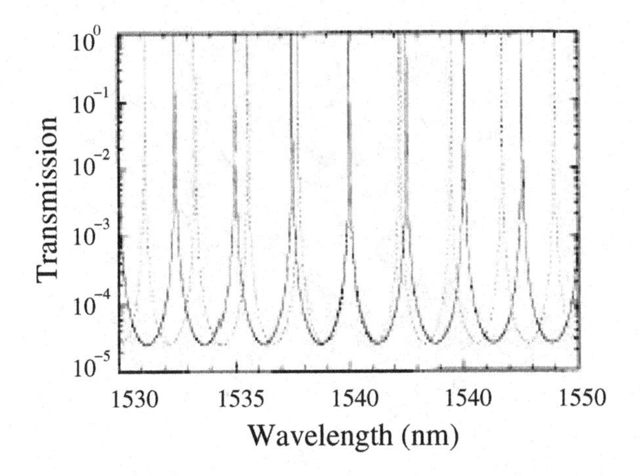

Fig. 3.24 The transmission spectra of two different rings. Adapted with permission from *IEEE Photo. Tech. Lett.* **14**, 5, pp. 601. Liu et al. (2002).

Table 3.2 Summaries of the features of monolithic tunable diode lasers.

Device	Integration	Tuning range	SMSR	References
Tunable DBR	Yes	>16 nm	>40 dB	[Debregeas-Silard *et. al.* (2002)]
SGDBR	Yes	>50 nm	>45 dB	[Shi *et. al.* (2002)]
SSG DBR	Yes	>62.4 nm	>35 dB	[Ishii *et. al.* (1996)]
GCSR DBR	Yes	>100 nm	>35 dB	[Rigole *et. al.* (1995)]
MEMS-VCSEL	No	> 31.6 nm	>35 dB	[Chang-Hasnain (2000)]
Others	No	>21.5 nm	<35 dB	[Uemukai *et. al.* (2000)]

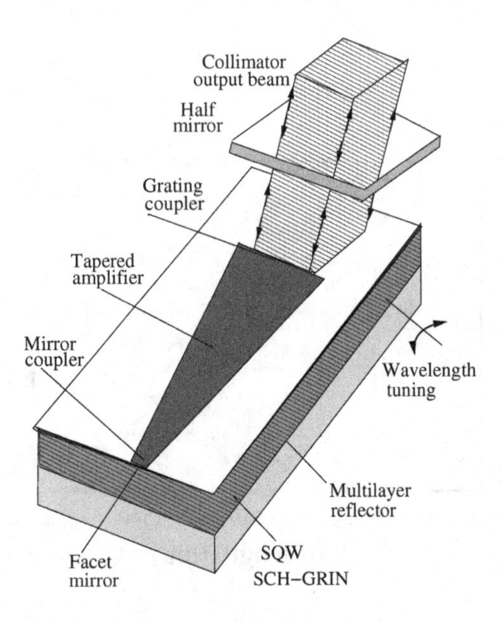

Fig. 3.25 Schematic of the tunable external-cavity laser. Adapted with permission from *IEEE Photo. Tech. Lett.* **12**, 12, pp. 1607. Uemukai et al. (2000).

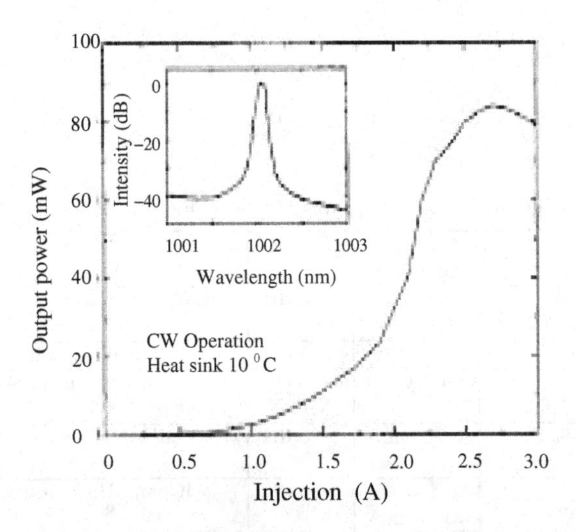

Fig. 3.26 Dependence of the output power through the half mirror on the injection current at $\theta = 11.5°$ under CW operation. The inset shows the lasing spectrum at I=1.8 A. Adapted with permission from *IEEE Photo. Tech. Lett.* **12**, 12, pp. 1608. Uemukai et al. (2000).

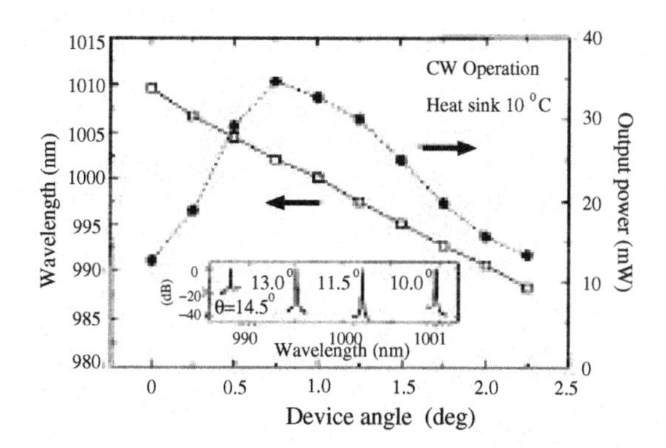

Fig. 3.27 Dependence of the lasing wavelength and the output power on the device angle for I=1.8 A, The inset shows the lasing spectra for various device angles. Adapted with permission from *IEEE Photo. Tech. Lett.* **12**, 12, pp. 1609. Uemukai et al. (2000).

Chapter 4

Elements for Tunable External Cavity Diode Lasers

4.1 Introduction

In previous chapters, we have studied the basic properties and the tunability of monolithic single-mode semiconductor laser. A number of other components are related in function, technology, and application. Tuning of diode lasers can be achieved by use of an external cavity as pointed out previously, in the following chapters, we are going to explore the properties of tunable external cavity diode lasers.

Recently, semiconductor diode lasers have been widely used in the scientific research and engineering technology. The application of diode laser to spectroscopy, the fundamental study of the interaction between matter and photons, especially in atomic physics have been extensively studied [Wieman and Hollberg (1991); MacAdam *et. al.* (1992); Ricci *et. al.* (1995); Fox *et. al.* (1997)], while the application to optical fiber telecommunications as local oscillator and coherent transmitter source have been soaring in recent years [Coldren (2003)].

The first experiment on laser diode coupled to external cavity has been demonstrated in 1964 [Crowe and Craig (1964)] soon after the first successful operation of diode laser. External cavity diode lasers (ECDLs) were studied in the early and late 1970s by a couple of research groups in Russia. Ludeke and Harris [Ludeke and Harris (1972)] reported the tunability of the cw radiation from GaAs injection laser in an external dispersive cavity over a range of 15 nm about the center wavelength of 825.5 nm at a temperature of 77 K. Single mode operation with cw output power as large as 17 mW with a linewidth of 350 MHz was observed. Fleming and Mooradian [Fleming and Mooradian (1981)] published the first paper in 1981 to study the property of ECDLs in details. Considerable works have been done in the early to mid-1980s at British Telecom Research Laboratory, motivated by the prospect of applying external cavity diode lasers as a transmitter and local oscillator in coherent optical telecommunications [Wyatt and Devlin (1983); Bagley *et. al.* (1990)]. In the same time, a lot of research have done at AT&T Bell Laboratories and Center National d'Etudes des Telecommunications (CNET) [Favre *et. al.* (1986)]. The end of 1980s and early 1990s witnessed growing interests in ECDLs as coherent radiation sources for spectroscopic research and in commercial fiber optic test and measurement equipment [Favre and Le Guen (1991)]. Nowadays, A widely tunable coherent sources are being developed for the applications in telecommunications and research by use of micro-electromechanical system to achieve the fast and accurate tuning of ECDLs.

A tunable external cavity diode laser could provide an alternative to monolithic semiconductor diode laser for accomplishing the widely tuning of diode laser. It is composed of a semiconductor diode laser with or without the antireflection coatings on one or two facets, collimator for coupling the output of the diode laser, and an external mode-selection filter. In general, the features of a diode laser in an external cavity can change greatly, depending on the external cavity length, the feedback level, optical power level, and the diode laser parameters [Lang and Kobayashi (1980) ; Sigg (1993)].

Monolithic semiconductor diode lasers are compact and robust, but they have numerous limitations for many applications. Solitary diode lasers are operating in multi-mode and exhibit broad linewidth, although they can be tuned by varying the temperature and current, the small tuning range can not meet many applications. However, external cavity diode lasers have very attractive linewidth as narrow as a few kilohertz, wide continuous tuning range of hundreds of nanometers, good single mode, and high stability.

In this chapter, we begin with an analysis of some basic but necessary elements of optics and electronics for ECDLs. Then, we move on to review the mode selection filters used as dispersive elements in the ECDL systems.

4.2 Optical coupling components

The external cavity diode lasers are basically composed of diode laser, beam collimator, dispersive elements, and good electronic control on the diode laser. In this section, we introduce elementary optical ingredients for ECDLs.

4.2.1 *Optical coating on laser facet*

A diode laser operation in an external cavity configuration can be optimized if the reflectance of the laser's output facet is extremely low. ECDLs are inherently coupled to the external cavity systems and are known to have their operating regions of stability and instability. Reduction of the reflectance of the output facet can dramatically improve the performance of the ECDLs: the range of the stable operating region (in temperature, current, feedback power) is enlarged; the usual output power is increased; and the laser's tuning range is broadened.

The requirement for strong external feedback is that the mirror losses of the solitary diode cavity are much greater than the combined mirror, mode selection filter, and coupling losses of external cavity [Zorabedian (1996)]. At a minimum, the solitary cavity loss should exceed the external cavity loss by at least 20 dB. For an external cavity configuration, in which the solitary and external cavities have one mirror in common, the requirement becomes $R_{facet} < 10^{-2} \times R_{ext}$, where R_{facet} and R_{ext} are the power reflectivities of the feedback-coupling facet and the external feedback optics, respectively.

For an ECDL system, an external feedback level of $R_{facet} \approx 0.1$ to 0.3 is typical value. Therefore, a rule of thumb is that the facet reflectance should 1×10^{-3} or less in order to maintain good ECDLs performance. Because the Fresnel reflectance of semiconductor-air interface is 0.31, some procedures to reduce the facet reflectance must be applied.

Almost all the commercial solitary diode lasers have already been coated on the laser's output facets. These coatings serve as two functions. Firstly, they protect the facets from degradation. Secondly, they adjust the facet reflectance to optimize the output power [Ladany *et. al.* (1977)]. Most commercial lasers can be greatly improved for operation in ECDLs configuration by reducing the reflectance of the output facet. In order to utilize the general-purpose lasers into the ECDL system, one has to find some special technologies to reduce the facet reflectance. One of the most commonly used methods is the deposition of a dielectric anti-reflection (AR) coating by single-layer.

For a plane wave incident at an interface between with index of refraction n_0 and a substrate of index n_s, a single dielectric layer of index $n_1 = \sqrt{n_0 n_s}$ and the thickness $t = \lambda/4n_1$ will reduce the reflectance to zero at a wavelength λ in air. Because of the finite lateral extent of the guided optical wave in the laser diode, the optimal coating design cannot be derived analytically as for a plane wave. The optimal film index and thickness values are $n_{opt} > n_m$ and that $t_{opt} > \lambda/4n_{opt}$, respectively, where n_m is the modal refractive index of the active region waveguide. A facet reflectance of 10^{-4} can be achieved with the film index and thickness tolerance of $\pm$ 0.02 and $\pm$ 2 nm, respectively. However, with careful process control or real time in situ monitoring the facet emission during coating, facet reflectance on the order of 10^{-4} can be obtained reproducibly with single-layer coatings [Wu *et. al.* (1992)].

Multi-layer dielectric coating is used to broaden the low-reflectance bandwidth and relax the thickness tolerances of the individual layers. Double-layer coating is applied in a high-low index sequence with the higher index layer in contact with the substrate. A maximally broad double-layer coating is obtained

$$n_1 = n_0(n_m/n_0)^{3/4}, \quad n_2 = n_0(n_m/n_0)^{1/4}, \tag{4.1}$$

and

$$t_1 = \lambda/4n_1, \quad t_2 = \lambda/4n_2, \tag{4.2}$$

where n_1 and t_1 are the index and thickness, respectively, of the inner layer, and n_2 and t_2 are those of the outer layer. This principle can be extended to three layers by incorporating a third quarter-wave layer with an intermediate index of refraction $n_3 = n_0(n_m/n_0)^{1/2}$ between the two layers specified above. Other index and thickness combinations for two-and three-layer antireflection coating are possible as well. Antireflection coatings with three dielectric layers in a low-high-low sequence of refractive indices have been used to relax the tolerance and broaden the low reflectance bandwidth.

The most widely used material for antireflection coatings on AlGaAs and InGaAsP facet is nonstoichiometric SiO_x, which can be deposited by two simple methods: First, thermally evaporated silicon monoxide. Second, e-beam- deposited HfO_2 and/or Al_2O_3. Either of these coatings can achieve reflectance below 10^{-3}, even on commercial lasers that already have been coated on their facets. Silicon monoxide is convenient because the equipment required for thermal evaporation is relatively simple. It also has useful property that the variation of the oxygen pressure in the coating chamber changes the oxygen composition (x) in the film, and hence the index of refraction from 1.6 to 2.0. With all the coatings, but with SiO_x in particular, the apparent index of refraction of the coating can change over time as the laser is exposed to and operated in air.

A detailed coating procedure has been described with SiO coating on standard commercial diode lasers [Boshier *et. al.* (1991)]. The first step is to expose the laser by removing the cap window of package. The coating is applied by evaporating a thin film onto the exposed output facet of the laser using a resistively heated thermal source. Of the materials which can be evaporated thermally, SiO is convenient and leads to an adequately low reflectivity ($\sim 1\%$) for a quarter-wave layer. Moreover, films of SiO have good stability and adhere well to cold substrates. The SiO is contained in a special boat which is baffled to prevent sudden eruption of the SiO onto the diode and is directly heated to 1250 K by passing a current of 250 A through it. The pressure in the coating chamber must be low enough to obtain the films of low reflectivity and good stability because the composition of the evaporated layer is affected by residual gases. Good results can be attained with a base pressure of 3×10^{-7} Torr.

The second step is to mount the diode in a holder that is approximately 30 cm away from the boat. Electrical feedthroughs in the vacuum chamber allows one to pass the current through the laser, and to monitor the light output using a photodiode. After heating and outpassing the boat, a shutter can be opened and the laser diode can be deposited. The correct thickness of the evaporated layer cannot be gauged with sufficient accuracy using a standard crystal monitor. Instead, one monitors the decrease of reflectivity directly by observing the corresponding decrease in the intracavity power of the laser. One has to increase the injection current to keep the laser just above the threshold, and when the reflectivity passes through the minimum value, the signal on the photodiode goes through a minimum, and thus the shutter should be closed. Since the power emitted from the coated front is considerably higher than emerging through the back facet into the photodiode, which is given by

$$\frac{P_c}{P_u} = \frac{(1 - R_c)\sqrt{R_u}}{(1 - R_u)\sqrt{R_c}}, \tag{4.3}$$

where R_c and R_u are the coated and uncoated reflectivities, respectively. Typically, the deposition takes three minutes, the boat is then allowed to cool down, the vacuum system is backfilled with dry nitrogen.

Figure 4.1 shows the output power versus the injection current for a gain-guided 670 nm laser diode before and after coating. If the coating is carried out as described above, it is typical to find no change in the

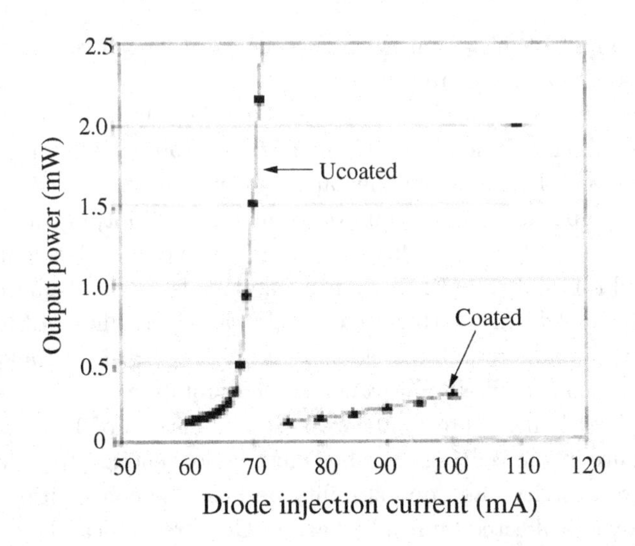

Fig. 4.1 L-I characteristic of a semiconductor laser diode before and after anti-reflection coating. Adapted with permission from *Opt. Commun.* **85**, pp. 356. Boshier et al. (1991).

power-current characteristic even after the diode has been operated for several months. Some lasers usually have a partial anti-reflection coating of Al_2O_3 (n=1.65). In this case, the minimum of reflectivity due to an additional layer of SiO is not so low and the tunability of the laser is adversely affected. Since almost all the commercial diode lasers have been coated on their facets, some of them work well in the ECDLs even without additional coating [Harvey and Myatt (1991); Atutov *et. al.* (1994); Andalkar *et. al.* (2000)]. Uncoated Fabry-Perot AlGaAs semiconductor lasers have been tuned over 105 nm in a grating-coupled external cavity with stripe contact single quantum lasers grown by molecular organic vapor chemical deposit [Mehuys *et. al.* (1989)].

Recently 80 GHz of mode-hop-free tuning from an external cavity diode laser without the need for additional coating on the diode facet has been reported [Petridis *et. al.* (2001)]. A wide continuous single mode tuning has been obtained by combining a simple electronic circuit and external cavity length in an appropriate ratio. The applicability of this technique to the most commonly available single-mode diode lasers without the need of antireflection coating makes it an attractive approach for the accomplishment of mode-hop-free tuning over a wide range.

Alternatively, lower facet reflectivity ($R = 10^{-5}$) can be achieved with an angled-facet structure [Rideout *et. al.* (1990)]. Angling the facet prevents reflected light from coupling back into the waveguide mode, that provides an inherently broadband reduction in facet reflectivity compared with AR coating. ECDLs based on a single-angled facet laser diode has been implemented and demonstrated for the first time [Heim *et. al.* (1997)] that wide tuning and high spectral purity can be achieved by use of the single-angled facet in the conventional external cavity configuration.

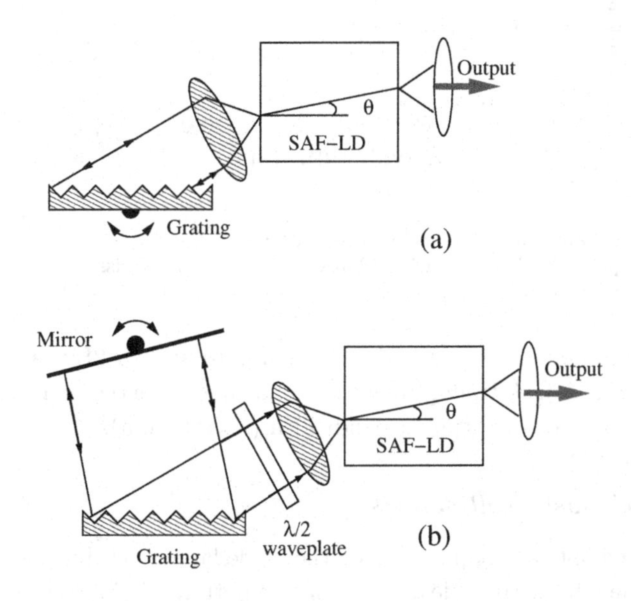

Fig. 4.2 Schematic diagram of single-angled facet laser diode (SAF-LD) external cavity diode laser. (a) Littrow configuration, and (b) Littman-Metcalf configuration. Adapted with permission from *Electron. Lett.* **33**, 16, pp. 1388. Heim et al. (1997).

Figure 4.2 illustrates Littrow and Littman-Metcalf configuration with InGaAs and InGaAsP single-angled facet laser diode mounted in, respectively. The lasing wavelength of both ECDLs configuration is adjusted by rotating the diffraction grating. The tuning range for the InGaAs ECDL is shown in Fig. 4.3. At a low bias current (I=90 mA), the tuning range is ~40 nm. A tuning range of 70 nm is achieved at a bias current of 190 mA corresponding to a 7% tuning bandwidth (λ=980 nm). The output power and spectral purity of the InGaAsP ECDL is measured, an output power of 13.5 mW (λ=1590 nm) has been obtained at a current of 160 mA with a threshold current 50 mA. The laser linewidth is measured by employing

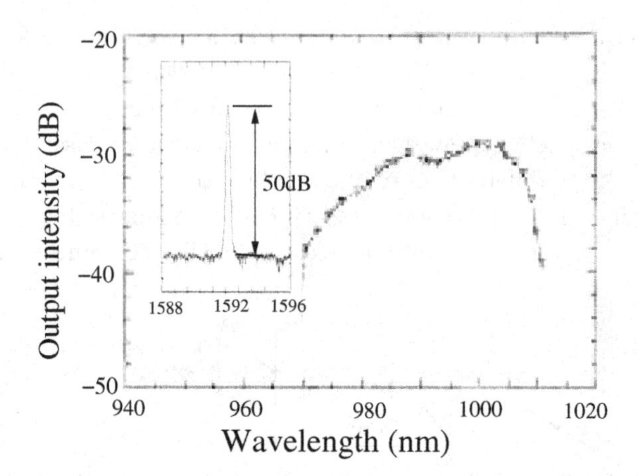

Fig. 4.3 Tuning range of single-angled facet external cavity diode laser at I=90 mA. Inset: Output spectra of SAF-LD at I=100 mA. Adapted with permission from *Electron. Lett.* **33**, 16, pp. 1388. Heim et al. (1997).

the self-delayed homodyne method and is found to be <50 kHz as shown in the inset of Fig. 4.3. A tuning range of 66 nm was observed at I=110 mA, that corresponds to a 4% tuning bandwidth (λ=1590 nm).

4.2.2 *Diode laser collimators*

The associated optics is required when the diode laser is used in the ECDLs because of the inherently wide angle radiation pattern of diode laser, arising from its small source dimensions. The radiation pattern can be as large as 90° in one dimension, which means that for efficient collection of laser radiation, the optics must have a large numerical aperture (NA). In addition, the highly coherent nature of the single-mode radiation dictates that the optics used with diode lasers must be diffraction limited if its full potential in terms of producing highly collimated beams or sharply focused spots is to be tapped. The wavelength, beam divergence, and beam ellipticity are the main diode laser properties that affect the design of the associated optics. The collimators must be relatively free of spherical aberration, with a maximum spot size of the order of the active-region cross-section dimensions. High optical throughput and negligible scattering are necessary to minimize cavity losses. If there is a window on the diode, window thickness, material, and separation from the diode laser surface must be taken into consideration as well.

Many different types of collimator lenses have been used to collimate the active area emission in ECDLs. Brief description of the most commonly used one and their properties are discussed as follows [Zorabedian (1996)].

i) Microscope objectives: These multi-element spherical lens system are available with numerical apertures as high as 0.8. To minimize the loss and spurious etalon effects, all external and internal optical surface should be antireflection coated. The selection of a specific set of lenses is based on an assessment of potential needs, it is presumed that the shorter focal length, high numerical aperture objectives would most often be used alone as beam collimator, while the longer focal length, lower numerical aperture lenses would be most useful for refocusing the beam. A large diameter, moderate numerical aperture lens is also included for producing a larger collimated beam for long distance applications. A beam expanding telescope that could be used in conjunction with one of the high numerical aperture collimating objectives is also included for long distance situations in which high throughput is needed.

ii) GRIN rod and silicon lenses: Rod lenses with a radically graded index of refraction are quite useful for ECDLs, but they have higher wavefront distortion than the best multiple- element systems, which probably reduces somewhat the maximum external feedback that can be obtained. The plano-plano versions have numerical apertures up to $\sim$ 0.45 NA, a plano-convex version has a 0.45 NA have been used [Mellis *et. al.* (1988)]. Singlet silicon lenses have lower spherical aberration for a given NA due to the high refractive index of silicon. Because silicon experiences strong absorption for wavelength less than $1.1\mu m$, these lenses are only useful for ECDLs working in the wavelength range of $1.3\mu m$ to $1.5\mu m$. Material dispersion may cause significant chromatic aberration and limit the tuning range otherwise that can be covered in the range of gain bandwidth.

iii) Spherical and ball lenses: Molded glass and plastic aspherics can be made with low wavefront distortion and are available with numerical apertures up to 0.55, glass is superior to plastic with respect to birefringence. Special high-index glasses reduce the severity of the aspheric curve needed to correct for spherical aberration, making the lenses easier to fabricate consistently. Molded aspherics are single-element lenses. Therefore, correction of chromatic dispersion is not possible. Dispersion in the lens material may limit the wavelength range that can be achieved without working distance adjustment. An ECDLs containing a molded- glass aspheric collimating lens has been reported [Harvey and Myatt (1991)]. Glass spheres can be used to couple the gain medium to waveguide or fiber-pigtailed external filters [Lau (1991); Lohmann and Syms (2003)]. However, the spherical aberration is too great to be useful for collimating in bulk optical cavity.

iv) Camera lenses: There are at least several published reports on the use of camera lenses as collimator in ECDLs. Heckscher and Rossi[Hechscher and Rossi (1975)] reported the use of a TV camera lens for intracavity collimating of a Littrow grating. The lenses gave only about 1% feedback when used with a grating, and it was concluded that spherical aberration was responsible for the poor performance since the lenses were not used in their intended geometry. Fleming and Mooradian successfully employed camera lenses in an ECDL system[Fleming and Mooradian (1981)].

4.2.3 *Beam expander and shaping*

There are two general and simple approaches to transform the elliptical diode laser beam cross-section emerging from a collimating lenses into a circle by utilizing an anamorpheric beam expander. One way is to use cylinder lenses, the other is to employ prisms [Zimmermann *et. al.* (1995)]. Beam expander which use cylinder lenses to circularize the beam usually do so in the form of a Gallilean beam expanding telescope. A cylinder lens has been used in ECDLs system to form a line illumination on a diffraction grating, this implements a degenerate resonator in one dimension and provides a high degree of angular misalignment tolerance with maintaining the high speed selectivity [Zorabedian and Trutna (1990)]. The advantages of using cylinder lens over prisms are that the beam is not displaced from the original centerline as it is expanded and two cylindrical elements can be adjusted to correct for any natural astigmatism in the diode output. Thus, the expander serves two purposes: shape and wavefront correction.

Unfortunately, the advantages described above are more than offset by a number of significant disadvantages. For large magnifications, a two element Gallilean telescope length becomes excessively long. An ideal anamorpheric expander should be made adjustable to accommodate the different elliptical ratio of the major and minor axis of the uncorrected output beam from diode lasers, which varies from diodes to diodes, cylinder lenses are also difficult to fabricate when quarter-wave quality or better is required.

The use of prisms allows one to overcome the disadvantages mentioned above in an ECDL system. The prisms are relatively easy to make with good transmitted wavefront and are easy to align [Zorabedian (1992)]. The most common prism configuration is the Brewster telescope as shown in Fig. 4.4. If the exit face of the wedge prism is made normal to the emerging beam, then the anamorpheric magnification, M of the prism is given by

$$M = H'/H = cosA/cosB, \tag{4.4}$$

the magnification of a pair of prisms is just the square of this value. A

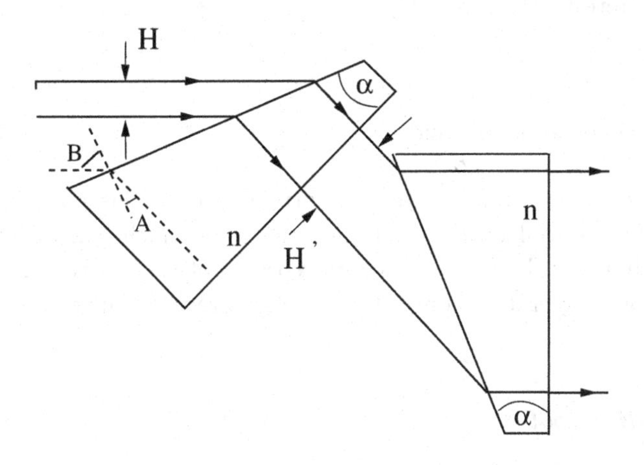

Fig. 4.4 Schematic diagram of magnification of a pair of prisms.

particular useful geometry is when the apex angle α is cut so that

$$\alpha = \pi/2 - \arctan\ n, \tag{4.5}$$

where n is the index of refraction of the prism material, this value of apex angle makes the maximum expansion of the pairs. The output beam is normal to the exit the prism when the incident angle equals the Brewster angle. The magnification of each prism is then equal to the index of refraction of the prism material, viz., M=n. It has the advantage of being compact and producing an exit beam parallel to the incoming beam. Difference in elliptical ratios can be accommodated by rotating the prism pair to different angles. The disadvantages are, at high magnification, the input faces of the prisms have to be antireflection coated and the beam is displaced off center from the incoming beam.

4.2.4 *Optical isolators*

ECDLs are very sensitive to the spurious optical feedback reverse-coupled through the output mirror. For very short cavity length, the feedback tolerance is as high as 20 dB, however the sensitivity increases with the cavity length. Isolation of at least 30 dB is typical used for external cavity length of 1 cm to 10 cm, up to 60 dB isolator is sometimes used. High isolation from back reflection is especially important when the output of the laser is being observed with a highly reflective instrument like a scanning Fabry-Perot interference.

4.3 Electrical control parts

Precise control over temperature and current of diode laser is highly desirable to an external cavity diode laser system, which can be achieved by accurately using the combination of Peltier cooler, sensitive thermistor, high precision temperature controller, and high precision current controller.

4.3.1 *Peltier cooler*

The principle of Peltier cooling is based on a basic concept named thermoelectric technology, which is known as Peltier effect. The effect occurs whenever electrical current flows through two dissimilar conductors, depending on the direction of current flow, the junction of the two conductors will either absorb or release heat. The cold side of the Peltier will simply absorb the heat and the flow of electrons will do the job of transferring the heat to the hot side to be dissipated. All this is only achieved with electron flow (electricity) from DC source, Fig. 4.5 shows the flow process.

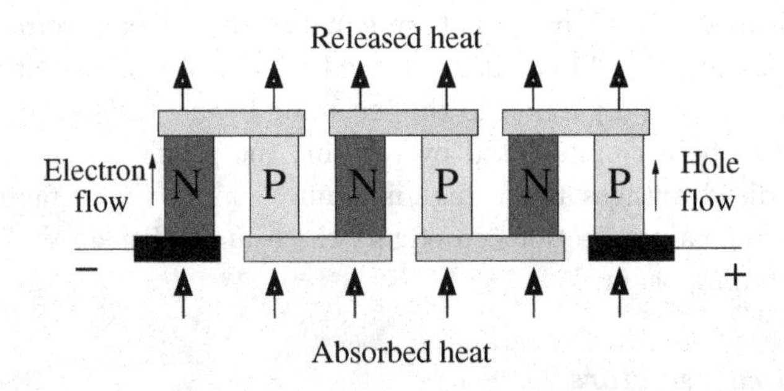

Fig. 4.5 Schematic diagram of Peltier cooler.

4.3.2 *Temperature sensor*

Thermistors are thermally sensitive resistors and have, according to type, a negative (NTC), or positive (PTC) resistance/temperature coefficient. Manufactured from the oxides of the transition metals - manganese, cobalt, copper and nickel, NTC thermistors are temperature dependent semiconductor resistors. NTCs offer mechanical, thermal and electrical stability, together with a high degree of sensitivity. PTC thermistors are also temperature dependent resistors manufactured from barium titanate. The other commonly used temperature sensor is AD590/592, which is a two-terminal integrated circuit temperature transducer, that produces an output current proportional to absolute temperature. For supply voltages between + 4 V and + 30 V, the device acts as a high impedance, constant current regulator passing 1 μA/K. This device allows one to read the temperature of diode laser easily.

4.3.3 *Temperature and current controller*

The electronics required to operate diode laser are relatively simply if one does not require precisely tuning the laser. Since the laser's output and frequency depend sensitively on the chip temperature and injection current, both of which are needed to be precisely controlled. Although it is adequate for some purposes to use only a single stage temperature control on the laser. In the long run, it is usually worth the same extra effect to add a second stage of temperature control on the base of the mount or on the box, that encloses the laser assembly [A1-Chalabi *et. al.*(1990); Talvitie *et. al.* (1997)].

For stable operation of diode lasers, it is necessary that both injection current and laser temperature be controlled. A detailed discussion of temperature control servo-loops has been given by Williams [Williams (1977)]. The basic idea is to employ a thermistor as one leg in a balanced bridge circuit, any voltage across the bridge is amplified and used to drive a the Peltier cooler, the reference voltage and the bridge resistors must have low temperature coefficients. One must carefully consider the thermal time delay and time constants in the electronics to achieve optimum performance. It is desirable to have the thermistor very close to the laser diode to accurately measure the laser's temperature, also very close to the Peltier cooler so that the time delay in the servo is small and the loop can have fast response and high gain.

These considerations indicate that a very small thermal mass should be used to support the laser. But the mount should also be mechanically stable and have a large thermal mass so that rapid temperature fluctuations do not perturb the laser's temperature. It is advisable to enclose the laser mount in some sort of hermetically sealed container to keep the dust out of the system and isolate the laser system from acoustic vibrations, thus to improve both thermal and opto-mechanical stability.

Temperature controller commonly employs a temperature sensor in a resistance bridge in a feedback loop, which regulates the power supplied to a heating/cooling element. Bridge feedback loop designs using either a constant or alternating reference voltage yield stability of better than $\pm 100\mu$K [Esman and Rode (1983); Dratler Jr. (1974)],

The most important requirement of current sources used for diode lasers is that they must be free of electrical transients that can be seriously damage the laser. The simplest laser current supply is just a battery and current-limiting resistor. A variety of good diode laser current sources are commercially available, for the ECDLs system, one needs to start with a low-noise current source and some protection against unwanted transients. An example of a good quality diode laser current source can be found in Refs. [Libbrecht and Hall (1993); Milic *et. al.* (1997)], where a new diode laser current controller was described, respectively, which features low current noise, excellent dc stability, and the capacity for high-speed modulation. While it is simple and inexpensive to construct, the controller compares favorably with the best presently available commercial diode laser current controllers. A constant current supply and a temperature control circuit have been developed for frequency-stable operation of laser diode [Bradley *et. al.* (1990)]. These instruments can stabilize laser diode injection current and temperature to better than ± 1 μA and ± 0.3 mK, respectively, over time periods exceeding 1 h. Others in literature [Cafferty and Thompson (1989)]. The primary consideration of design is to have good stability, low noise, and modulation capability.

4.3.4 *Piezoelectrical transducer (PZT)*

The last electrical part worth mentioning is piezoelectrical transducer (PZT). Piezoelectric transducers are solid state (ceramic) actuators, that convert electric energy directly into mechanical energy by motion of extremely high resolution. In the ECDLs, we use the PZT to alter the cavity length, thus to cause the tuning of the laser frequency with electrical control. Usually one uses the PZT stack or disk to provide displacement of about several micrometers by applying the scanning voltage to PZT.

4.4 Mechanically tuned mode-selection filters

In order to select the single mode from many modes of diode lasers, A number of mode selection filters are used in ECDLs system, which ideally have a bandwidth narrower than the longitudinal mode spacing of the external cavity, and zero insertion loss at its peak. Various mode-selection filters are considered in this section.

4.4.1 *Diffraction gratings*

Diffraction gratings are the most common used filter in the ECDLs, a diffraction grating can consist of a periodic variation of thickness in a medium of constant refractive index or a periodic variation of refractive index in a medium of constant thickness [Palmer (2002)]. The first type of diffraction grating is a surface relief grating, we mainly introduce this kind of grating. The second type of grating is a volume phase grating, this kind of fiber Bragg grating will be briefly discussed.

Figure 4.6 schematically shows a cross section of a surface relief grating. The period d is generally referred to as the grating spacing, or the groove spacing, the inverse of this period is called spatial frequency, or groove density, ρ. When a beam of light is incident on a grating, each groove produces a diffracted wavelet. For each wavelength component in the incident beam, the constructive interference of the diffracted components from each groove occurs, this means the interference condition is fulfilled when the path difference is equal to multiples, diffraction orders q, of the wavelength of incident light. This gives rise to the grating equation:

$$q\lambda = d(\ sin\alpha + \sin\beta), \tag{4.6}$$

or

$$q\rho\lambda = \sin\alpha + \sin\beta, \tag{4.7}$$

where λ is the wavelength of incident light, α and β are the incident and diffracted angles, respectively. A special but very common case is that in which the light is diffracted back toward the direction from which it came, i.e., $\alpha = \beta$, this is called the Littrow configuration, for which the grating

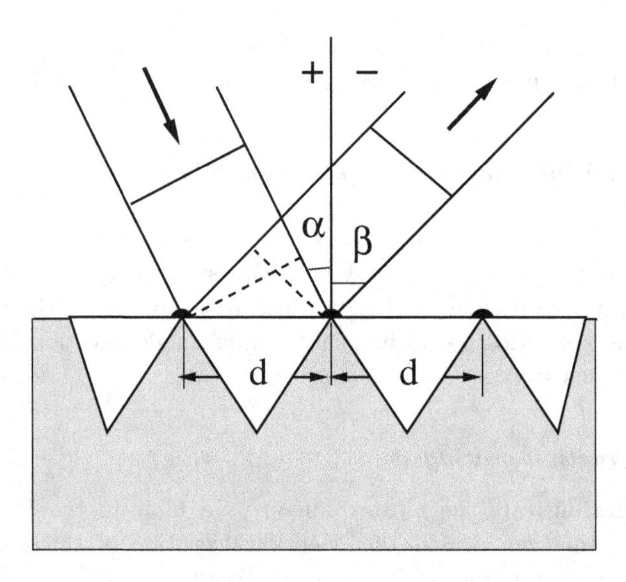

Fig. 4.6 Geometry of diffraction for plane wavefronts.

equation becomes

$$q\lambda = 2d \sin \alpha. \tag{4.8}$$

The primary purpose of a diffraction grating is to disperse light spatially by wavelength. Dispersion is a measure of the separation from diffracted light with various wavelengths. By differentiating with respect to the output angle, one obtains the angular dispersion

$$D = \frac{\partial \beta}{\partial \lambda} = \frac{q}{d \cos \beta} = \rho q \sec \beta. \tag{4.9}$$

Substitution of the grating equation in Eq. (4.9) yields the following equation for the angular dispersion

$$D = \frac{\partial \beta}{\partial \lambda} = \frac{\sin \alpha + \sin \beta}{\lambda \cos \beta}. \tag{4.10}$$

This becomes even clear when we consider the Littrow configuration, that is

$$D = \frac{\partial \beta}{\partial \lambda} = \frac{2}{\lambda} \tan \beta. \tag{4.11}$$

In the grazing-incidence configuration, also called Littman-Metcalf configuration [Littman and Metcalf (1978)], one of the other popular ECDLs schemes, the intracavity beam passes through the grating twice. The diffracted laser beam from the second pass is retroflection of the incident laser from the first pass. Thus the angular dispersion of the retroreflected light is twice that of the laser diffracted on one pass. The dispersion in this case is

$$D = 2\frac{\partial \beta}{\partial \lambda} = \frac{4}{\lambda}\tan\beta. \tag{4.12}$$

The wavelength resolution of a grating-tuned external cavity is determined by the angular dispersion multiplied by the acceptance angle for coupling back into the diode laser. The angular dispersion can be a kind of figure of merit for an ECDL system.

The resolving power of a grating is a measurement of its ability to separate adjacent spectra lines of average wavelength λ. It is usually described by

$$R = \frac{\lambda}{\Delta\lambda}, \tag{4.13}$$

here $\Delta\lambda$ is the limit of resolution, the difference in wavelength between two lines of equal intensity that can be distinguished. The theoretical resolving power of a planar grating is given by

$$R = qN = \frac{Nd(\sin\alpha + \sin\beta)}{\lambda}, \tag{4.14}$$

where q is the diffraction order and N is the total number of grooves illuminated on the surface of the grating. The maximal resolving power can be $R_{max} = 2W/\lambda$, where $W = Nd$ is the width of thè illuminated region on the grating. The maximum condition corresponds to the grazing Littrow configuration, i.e., $\alpha \approx \beta$ (Littrow) and $\alpha = 90°$ (grazing).

The wavelength resolution is obtained by dividing the angular spread of the beam waist (beam divergence) at the grating by the angular dispersion, that is

$$\Delta\lambda = \frac{\Delta\beta}{D}. \tag{4.15}$$

The beam divergence can be obtained by

$$\Delta\beta = \frac{\lambda}{\pi w}, \tag{4.16}$$

where w is the Gaussian beam of radius. Therefore, the wavelength resolution for the Littrow and Metcalf-Littman cases are, respectively,

$$\Delta\lambda = \frac{\Delta\beta}{D} = \frac{\lambda^2}{2\pi w \tan\beta}, \tag{4.17}$$

and

$$\Delta\lambda = \frac{\Delta\beta}{D} = \frac{\lambda^2}{4\pi w \tan\beta}. \tag{4.18}$$

The above expression can be simplified by introducing the filled depth of the grating, which is the projection of the illuminated region of the grating onto the optical axis of the cavity and given by

$$L_g = 2w \tan\beta, \tag{4.19}$$

Eqs. (4.17) and (4.18) turn out to be

$$\Delta\lambda = \frac{\lambda^2}{\pi L_g}, \tag{4.20}$$

and

$$\Delta\lambda = \frac{\lambda^2}{2\pi L_g}. \tag{4.21}$$

The diffraction efficiency (DE) of a diffraction grating is simply the ratio of the power in the diffracted beam to that in the incident beam. In addition to the fundamental diffraction efficiency of the grating itself, the overall diffraction efficiency includes the reflection losses at the entrance, exit surfaces, any internal scattering and absorption losses. The term insertion loss (IL) is expressed in decibels (dB), it is defined as:

$$IL = 10 \times \log(1/DE)\ (dB). \tag{4.22}$$

The diffraction efficiency of a grating is wavelength dependent and the degree to which it is wavelength dependent will determine the grating's usefulness as a dispersive element in the ECDLs system. The diffraction efficiency of a grating is also generally polarization dependent. That is, the diffraction efficiency of the grating will depend on the polarization direction of the incident beam relative to the plane of incidence. This polarization dependence results in a difference in insertion loss that is defined as the polarization dependent loss, or PDL, like IL. PDL is expressed in dB and is just the difference in insertion loss for two orthogonal polarizations.

PDL is a concern in an ECDL system, for laser beam polarization perpendicular to the grating rulings, high efficiency is expected, while for beams parallel, the efficiency can be significantly reduced. In the Littrow configuration, the number of grating lines covered by the laser mode is constrained by the focal length, numerical aperture of the collimating lens, the orientation of laser spatial mode, the line spacing, and corresponding cutoff wavelength of grating. Since the beam from a laser waveguide is normally polarized in the direction parallel to the junction. When the laser asymmetric mode is incident on a diffraction grating, there is a compromise between high resolution and high grating efficiency. In some cases it can be advantageous to use a half-wave plate between the laser and grating to decouple the polarization from the spatial mode orientation.

4.4.2 *Fiber Bragg gratings*

A fiber grating may be manufactured by exposing a length of the fiber core to a nearly sinusoidal varying intensity of UV light. The fiber is exposed to the interference pattern of the two laser beams or to that created by a laser beam traversing a phase mask, this UV light intensity exposure then impose a periodic index along the length of the fiber core by creating a corresponding periodic concentration of the glass defects.

Fiber Bragg grating generally refers to the device with refractive index maximum spacing on the order of 1/2 or 1 times the wavelength of the guide mode. These grating couple the forward-propagation core mode to back- propagation guided modes. As shown in Fig. 4.7, the grating reflects light of wavelength λ_{Bragg}, such that

$$\lambda_{Bragg} = 2n_{eff}\Lambda, \tag{4.23}$$

where n_{eff} is the mode effective index of the guided core mode and Λ is the periodicity of the index grating. In addition, the use of a fiber grating enables the wavelength of operation to be selected from the broad optical gain bandwidth of a single facet antireflection coated diode lasers. The low coupling loss, narrow linewidth of 50 kHz, and simplicity of packaging a fiber external cavity with Bragg reflector in an ECDL system has been demonstrated [Bird *et. al.* (1991)].

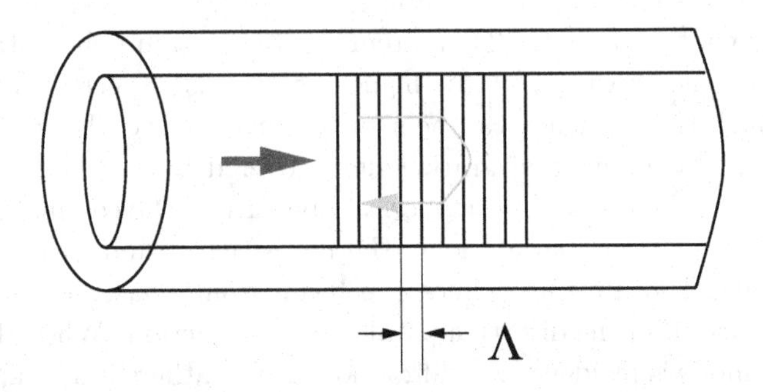

Fig. 4.7 Schematic of fiber Bragg grating.

4.4.3 *Fabry-Perot interferometer and bandpass interference filter*

The Fabry-Perot interferometer consists of two plane or spherical mirrors with amplitude reflectivities r_1 and r_2 separated by a distance d, it contains a medium of refractive index n. The power transmission is given by

$$T = \frac{(1-\sqrt{r_1})(1-\sqrt{r_2})}{[1-(r_1r_2)^{1/4}]^2 + 4(r_1r_2)^{1/4}\sin^2\phi}, \tag{4.24}$$

where $\phi = 2\pi\nu nd/c$, the maximum transmission occurs at $\phi = q\pi$, where q is positive integer. Therefore, the frequencies ν_q of these maxima are obtained by

$$\nu_q = \frac{qc}{2nl}, \tag{4.25}$$

the frequency difference between two consecutive maxima is called free spectral range of the FPI, we have

$$\Delta\nu_{fsr} = \frac{c}{2nl}. \tag{4.26}$$

If we define the reflectivity R associated with the intensity as $R=r^2$. The half-power bandwidth (HPBW), or its 3dB bandwidth, is given by

$$\Delta\nu_c = \frac{c}{2nl} \cdot \frac{1-\sqrt{R_1R_2}}{\pi(R_1R_2)^{1/4}}. \tag{4.27}$$

The very important performance parameter characterizing a FP filter is the finesse, the ratio of FSR to HPBW, which express the sharpness of the filter relative to the repeated peak, the finesse is defined as

$$F = \frac{\Delta\nu_{fsr}}{\Delta\nu_c} = \frac{\pi(R_1R_2)^{1/4}}{1-(R_1R_2)^{1/2}}, \tag{4.28}$$

it is typically much greater than 1.

The simplest bandpass filter is a very thin Fabry-Perot interference, the gap is a thin layer of dielectric material with a half-wave optical thickness. The high reflectors are normal quarter-wave stacks with a broadband reflectance peaking at design wavelength. The entire assembly of two quarter-wave stacks is applied to a single surface, the simplest bandpass interference filters are sometimes called cavities. Two or more such filters can be deposited on top of the other, the advantages of multi-cavity filters are steeper band slopes and the improved near-band rejection.

A common characteristic of single- and multi-layer dielectric coatings and interference filters is that the transmittance and reflectance spectra shift to shorter wavelengths as they are tilted from normal to oblique incidence. In terms of the external angle of incidence α, it can be shown that the wavelength of peak transmittance at small angles of incidence is given by

$$\lambda = \lambda_{max}\sqrt{1-(n_o/n)^2\sin^2\alpha}, \tag{4.29}$$

where n_o is the external medium refractive index and n is the spacer effective refractive index, which depends on wavelength, film material, and order number. Bandpass filters can be made with HPBW 2 nm or less in the near infrared and 1 nm in the visible. The peak transmission can be as high as 50% to 70% [Boshier *et. al.* (1991)].

4.5 Electronically tuned mode selection filters

4.5.1 *Liquid crystal spatial light modulator*

A liquid crystal spatial light modulator (LC SLM) can be thought of as a pixellated variable wave plate, with each pixel acting as an individual waveplate, it provides an ability to fast modulate the phase and amplitude of light with very high resolution. As such, the SLM can act in two different modes, in an amplitude modulation mode and a phase modulation mode. The modulation is achieved by feeding the desired signal to the SLM either optically (signal is imaged through a LCD onto the SLM) or electrically (pixels are addressed individually).

4.5.1.1 *Liquid crystal*

Liquid crystals are material which shares many features attributable to liquids, together with molecular ordering properties normally associated with crystals in particular. The molecular ordering of any particular liquid crystal is usually temperature related (thermotropic), the particular state of the liquid crystal is termed as a mesophase.

Liquid crystal molecules are generally rod shaped and each molecule has an associated net dipole moment as a whole. It is this diploe moment that leads to the spontaneous molecular ordering within the material, and which allows this ordering to be changed upon application of an external electric field. Liquid crystals are commonly classified into three groupings - nematic, cholesteric and smectic, and of particular interest is the nematic liquid crystal.

Nematic liquid crystals posses a dielectric anisotropy, so that the dielectric permittivity ε varies with respect to the angle relative to the director orientation. As a matter of fact, nematic liquid crystals are uniaxial and birefringent. Therefore, linearly polarized light propagating through the medium experiences two different index of refraction according to whether the dipole moment is parallel or perpendicular to the director orientation. The dielectric anisotropy $\delta\varepsilon$ is defined as

$$\delta\varepsilon = \varepsilon_{\parallel} - \varepsilon_{\perp}. \tag{4.30}$$

The birefringence associated with nematic liquid crystals is extremely large. If n_e denotes the index of refraction as seen by light polarized such that the electric field vibration is parallel to the director, and n_o the corresponding index for light polarized perpendicular to the director, the birefringence is defined as

$$\delta n = n_e - n_o. \tag{4.31}$$

The most straightforward electro-optical effect utilizing nematic liquid crystals is known as 'field induced birefringence'. This effect has the distinct advantage that either amplitude or phase modulation can be achieved by suitable orientation of a polarizer-analyzer pair as described below.

Both phase modulation and amplitude modulation strongly depend on the voltage controllable molecular tilt, and consequent variation of n_e with depth for their success. Consider a linearly polarized beam of light incident at an angle α to the director at the surface of the cell. The optical phase of the beam component traversing the cell perpendicular to the surface director is given by

$$\varphi_o = \frac{2\pi\delta n_o l}{\lambda}, \tag{4.32}$$

where l is the cell thickness, λ is incident wavelength. The optical phase of the beam component polarized parallel to the surface director is, on the other hand, given by

$$\varphi_e = \frac{2\pi}{\lambda}\int_0^l n_e(z, V)dz, \tag{4.33}$$

where z denotes distance through the cell, V is the voltage across the cell. As n_e is both a function of depth and applied voltage. As such, the parallel, homogeneous cell configuration utilizing nematic liquid crystal (of positive dielectric anisotropy) acts as a wave plate of continuously variable retardance. Eq. (4.33) is of particular interest for the integral relatively insensitive to variations in the cell thickness.

4.5.1.2 *Amplitude modulation*

Amplitude modulation is achieved by ensuring the incident light field polarized at $\pi/4$ to the surface director, the components of the beam perpendicular and parallel to the surface director have equal intensity. Upon emerging from the cell, the relative phase difference between both components, the ordinary and extraordinary rays, is given by

$$\Delta\varphi = \frac{2\pi}{\lambda}[\int_0^l n_e(z, V)dz - n_o l]. \tag{4.34}$$

The output intensity emerging from an analyzer at $\pi/2$ to the first polarizer in this situation is given by

$$I = I_0 \sin^2\frac{\Delta\varphi}{2}, \tag{4.35}$$

where I_0 is the maximum transmitted intensity, so that an optical phase difference of $\Delta\varphi = 2q\pi$, q integer, results in a minimum of transmitted light. In practice the voltage is selected to obtain this state, and may also be adjusted to obtain a maximum cell transmittance. A cell which gives maximum transmission is said to be in an 'ON' state (with applied voltage

V_{ON}) and the other giving minimum transmission as being in an 'OFF' state (with applied voltage V_{OFF}).

4.5.1.3 *Phase modulation*

Pure phase modulation is fulfilled by ensuring the input light beam polarized parallel to the surface director. As such only the voltage-dependent extraordinary refractive index of the liquid crystal causes the phase delay as the beam passes through the cell. Increasing the voltage across the cell causes molecular tilt towards the cell normal, reducing n_e on average, and thus decreasing the optical path of the material. By this mechanism, a spatial light modulator with variable pixel voltage can advance or retard the phase of the light emerging from a pixel relative to another that from another pixel held at a fixed voltage.

Phase modulation where the difference in phase between light emerging from two pixels is either 0 or π radians is achieved as follows. Consider one pixel of a spatial light modulator (or a liquid crystal test cell) which has been set up to perform amplitude modulation, and the cell is in an 'OFF' state. The difference optical path length through the cell for the orthogonal polarization states of the incident light beam is

$$\varphi_e(OFF) - \varphi_o = 2q\pi, \tag{4.36}$$

where q is an integer. Now consider a neighboring pixel (or another test cell) in an 'ON' state. The optical path length difference between the orthogonal beam components is now

$$\varphi_e(ON) - \varphi_o = (2q+1)\pi. \tag{4.37}$$

Note that φ_o is independent of the applied voltage in each case. If the input polarizer is now rotated so that the incident polarization lies completely parallel to the surface director, the difference in optical path for the e- rays in pixels which were 'ON' and 'OFF' in amplitude mode is

$$\varphi_e(ON) - \varphi_e(OFF) = \pi, \tag{4.38}$$

so that if the analyzer is removed from the system, pure binary phase modulation of 0 or π radians between pixels occurs. This method of phase modulation was chosen to achieve pure phase modulation using nematic liquid crystals.

4.5.2 *Birefringence filter*

More than 50 years ago, B. Loyt introduced new types of optical filters called birefringent filters, commonly referred to as Brewster angled Loyt filters [Loyt (1933)]. Such filters take advantage of the phase shifts between orthogonal polarization to obtain narrow band filters. It requires birefringent wave plates introducing the phase retardation between the two orthogonal components of a linearly polarized light that correspond to the fast and slow axes of the birefringent material.

Loyt filter comprises of an alternating stack of N uniaxial birefringent plates separated by polarizers. The thickness of plates varies in a geometrical progression d,2d,4d,...,2^{N-1}d. The transmission axes of the polarizers are all aligned. The light propagates in a direction perpendicular to the c axis of each of the plates. Transmission through each segment will vary sinusoidally with a maxima at wavelength for which the retardation of the plate is an integral multiple of 2π. For a plate of thickness d, the free spectral range $\Delta\lambda_{FSR}$ between successive maxima is approximately by

$$\Delta\lambda_{FSR} \approx \frac{\lambda}{d} \frac{1}{(\partial\delta n/\partial\lambda) - \delta n/\lambda}. \tag{4.39}$$

For each segment, the separation between transmission maxima and the full width at half maximum (FWHM) of one of the maxima is inversely proportional to the plate thickness. Thus, the resulting transmission for the entire stack will consist of narrow bands having the FWHM of the thickest plate and separated by the free spectral range of the thinnest plate. Electrically tuned birefringent filters can be realized using liquid crystal cells as the birefringent plates as described previously.

A six-stage birefringence filter placed in an external cavity of an AlGaAs diode laser has been used to scan the laser electronically over 10.3 nm [Andrews (1991)]. Continuous electronic tuning of the single mode over the free spectral range of the external cavity 182 MHz was also demonstrated by use of a variable-phase plate. Nematic liquid-crystal cells are used to attain both tuning of functions with applied voltage less than 2 V.

The laser schematic is shown in Fig. 4.8(a). The diode laser was placed at one end of an external cavity consisting of a collimator, and a partially transmitting mirror with 90% reflectivity. An intracavity six-stage birefringent filter is used for fine tuning the cavity resonance frequency. By placing the wavelength-dependent loss of the birefringent filter in the laser external cavity, the maximum net gain, and consequently the laser wavelength, is the

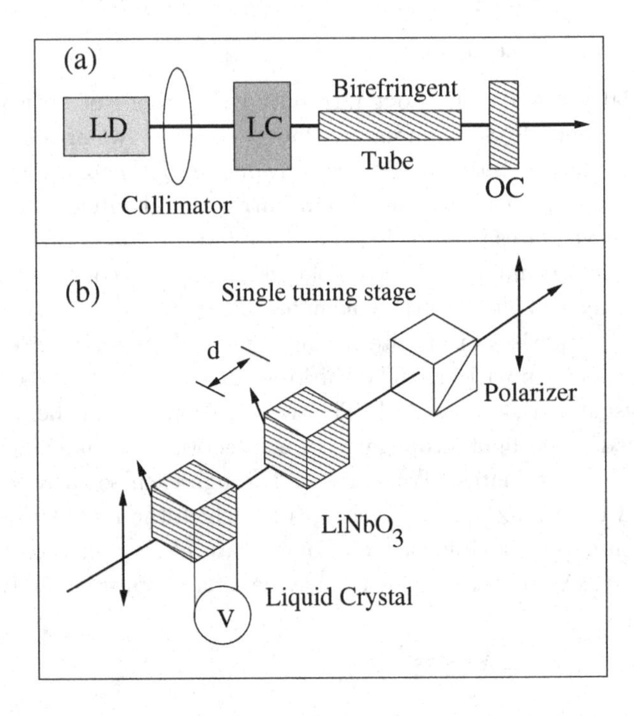

Fig. 4.8 (a) Experimental setup for external-cavity laser. The birefringent filter consists of six stages. (b) A single stage of the filter, each filter stage varies in the thickness d of the $LiNbO_3$ crystal, LD: lase diode; LC: liquid crystal; OC: output coupler. Adapted with permission from *Opt. Lett.* **16**, 10, pp. 732. Andrews (1991).

transmission maximum of the filter. The variable-phase plate permits fine wavelength tuning by changing the effective optical length, which in turn changes the resonance frequency of the external cavity Fabry-Perot modes. Each stage of the birefringent filter, as shown Fig. 4.8(b), consists of a fixed birefringence $LiNbO_3$ crystal, a variable birefringent plate, and a thin-film polarizing cube. The polarizer transmission for each filter stage is parallel to TE polarization of laser diode. The voltage to each of the liquid-crystal cells originated from a single square-wave generator operating at 1 kHz. A parallel voltage divider network supplies an adjustable voltage of as much as 5 V to each of the cells.

The variable-birefringence plate in each of the six filter stage and the variable-phase plate are liquid-crystal cells of the same design. The cell has a nominal spacing of 6.5 μm and contained the nematic liquid crystal. The surface alignment for the liquid crystal is anti-parallel. This results in a refractive index that varies with the voltage for light polarized parallel to the alignment direction and a voltage-independent refractive index for light polarized perpendicular to the alignment direction. In each filter stage the alignment direction of the liquid-crystal cell is oriented at 45 ° to the transmission axis of the polarizers, which leads to a voltage-dependent birefringence. For the variable-phase plate, the alignment direction is parallel to the polarizer transmission axis, then resulting in a voltage-dependent refractive index. The ECDL containing an antireflection-coated lithium tantalate electro-optic crystal has been demonstrated to provide a rapid frequency tuning [Greiner *et. al.* (1998)]. The crystal provides continuous frequency over approximately one external-cavity spacing of ~4 GHz with the 3 cm-optical-path-length external-cavity.

4.5.3 *Acousto-optic tunable filter*

Principle of operation of acousto-optic tunable filters (AOTFs) resemble interference filters in operation. An acousto-optic tunable filter is a solid state, electronically tunable bandpass filter which is based on acoustic diffraction of light by the acousto-optic interaction in an anisotropic medium. The filter allows one to select and transmit a single wavelength from the broadband semiconductor laser.

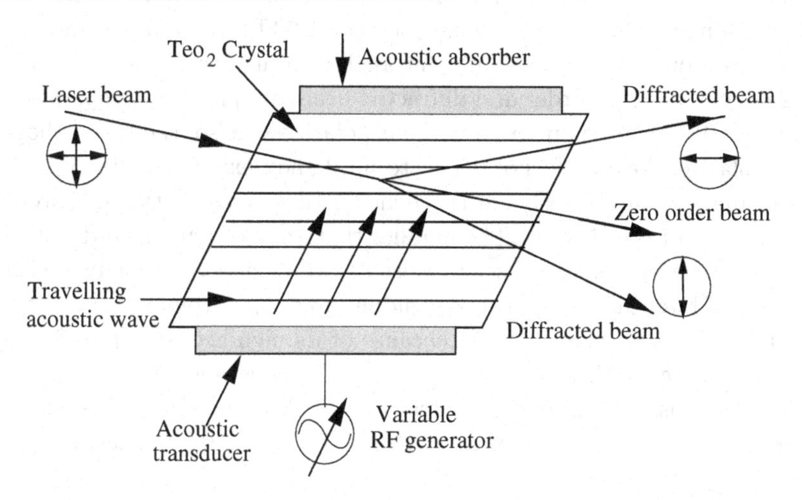

Fig. 4.9 Schematic diagram of acousto-optic tunable filters.

As shown in Fig. 4.9, the device consists of a piezoelectric transducer

(PZT) bonded to a birefringent crystal. When the transducer is excited by an applied radio frequency (RF) signal, acoustic waves are generated in the medium. The propagating acoustic wave produces a periodic modulation of the index of refraction, this provides a moving phase grating that, under proper conditions, will diffract portions of an incident beam. For a fixed acoustic frequency, only a limited band of optical frequencies can satisfy the phase-matching condition and be cumulatively diffracted. The RF frequency applied to AOTFs transducer controls the transmitted laser wavelength, the center of optical passband is changed accordingly so that the phase-matching condition is maintained. The RF amplitude applied to AOTFs transducer controls the transmitted laser intensity level, this is a unique feature provided by the AOTFs, a AOTF has a fast response time of microsecond scale and exhibits a high extinction ratio.

AOTF devices fall into two categories in terms of their configurations (see Fig. 4.9). In a quartz collinear AOTF, the incident light, the diffracted filtered light and the acoustic wave all interact collinearly in a birefringent crystal. As a result of the acousto-optic interaction, part of the incident light beam within the filter spectral passband is coupled to the diffracted light beam. The polarization of the incident light beam is orthogonal to that of the diffracted light beam. Because of the zero-order beam and the diffracted beam are collinear, polarizers must be used to separate them.

In a tellurium dioxide (TeO_2) noncollinear AOTF, the acoustic and optical waves propagate at quite different angles through the crystal. In this configuration, the zero-order and diffracted beam are physically separated, so that the filter can be operated without polarizers. Also, the two orthogonally polarized beams do not separate until they exit from the crystal, and the angle of diffracted beam is absent for the change in the first order with a change of wavelength. This implies that only a single fixed detector is necessary during a spectral scan. Most AOTF devices are designed with two types of birefringent crystals depending upon operational wavelength. TeO_2 is preferred AOTF material because of its high acousto-optic figure of merit. The crystal, although useful in the visible and infrared region up to 4.5 μm, is not suitable for ultraviolet applications due to its short-wavelength transmission cutoff at 350 nm. For ultraviolet spectroscopy, crystalline quartz is used.

The peak wavelength of the transmission passband λ_p is given by

$$\lambda_p = \frac{v \delta n}{f_r} \gamma, \tag{4.40}$$

where v is the acoustic velocity, f_r is the acoustic frequency, and δn is the crystal birefringence, and γ is a dimensionless parameter, whose value depends on the orientations of the various beams with respect to the crystallographic axis. The passband width of an acousto-optic filter is given by λ_p

$$\lambda_{FWHM} = \frac{\lambda_p^2}{l\delta n}\gamma, \tag{4.41}$$

where l is the acousto-optic interaction length, sub-nanometer resolution in the visible and a FWHM of ~1 nm at around 1.3 μm have been achieved.

An electrically tunable semiconductor laser system was demonstrated [Hidaka and Nakamoto (1989)] using an acousto-optic (AO) device and commercially available semiconductor laser, rapid scanning of laser oscillation frequency was obtained using frequency modulation of the driving RF power for the AO device. The experimental setup is shown schematically in Fig. 4.10. The laser are free running with wavelength 788 nm without

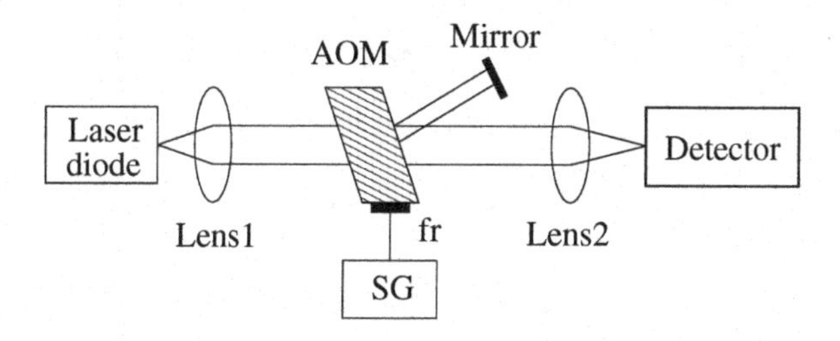

Fig. 4.10 Experimental setup of electrically tunable laser diode with acoustic-optic modulator (AOM). SG: signal generator; f_r: frequency of radio frequency. Adapted with permission from *Electron. Lett.* **25**, 19, pp. 1320. Hidaka and Nakamoto (1989).

antireflection coating, the output beam is collimated by lens L_1, the collimated beam is diffracted by an acousto-optic device made from TeO_2. The oscillation wavelength is given by Eq. (4.40), here γ is diffracted angle.

When a mirror reflects the diffracted beam vertically, the light returns to the laser diode, resulting in the extra gain at wavelength λ_p, one can tune the laser oscillation wavelength λ_p by tuning the radio frequency f_r. Coquin and Cheung [Coquin and Cheung (1988)] have demonstrated electronic tunability of 35 nm in an external cavity GaAs semiconductor laser by using a pair acousto-optic tunable filter (AOTF) and an AO modulator inside the cavity to select the wavelength. This allows fast and accurate selection of the lasing wavelength by varying only the drive frequency of the AO devices and in times of approximately 10 μs or less.

Chapter 5

Systems for Tunable External Cavity Diode Lasers

In the previous chapter, we considered commonly-used components for developing external cavity diode lasers. In this chapter, we are in a position to examine the system of external cavity diode lasers. General external optical feedback effects on diode lasers are concerned with a three-mirror laser cavity model, and in the case of steady and dynamic state analysis with various feedback strengths. Spectral characteristics of ECDLs are presented in terms of output power, single mode tunability, linewidth, and wavelength dependence of temperature. Systems of the tunable external cavity diode laser are introduced with a number of external cavity designs. Finally we deal with the alignment procedures for single-mode operation and the configuration for mode-hop suppression in external cavity diode laser systems.

5.1 Optical feedback in external cavity diode lasers

The sensitivity of the output intensity of a diode laser to both the amplitude and phase of external optical feedback is well documented [Eliseev *et. al.* (1969); Salathe (1979); Glasser (1980); Olsson and Tang (1981); Liu *et. al.* (1984); Sivaprakasam *et. al.* (1996)]. There has been substantial interest in the spectral characteristics of the emission under various external feedback, including the introduction of external-cavity dispersive elements. The effects of optical feedback on the behavior of diode laser are complicated and have been studied in the early 1980s [Lang and Kobayashi (1980)]. It has been shown that the dynamic properties of injection lasers are significantly affected by the external feedback, depending on the interference conditions between the laser field and the delayed field (returning from the external cavity). The essence of the optical feedback method is to increase the quality factor (Q) of laser's resonator, therefore narrowing the linewidth and stabilizing the laser's wavelength. Petuchowski et al. has reported [Petuchowski *et. al.* (1982)] the behavior and spectral features of a constricted heterojunction injection laser in a regime, where the lasing characteristics is dominated by a cavity composed of two external mirror. The Fox-Smith configuration was shown to limit the laser's multimode emission to a single longitudinal mode with a linewidth of less than 200 kHz.

In this section, we attain the threshold and phase conditions for external cavity lasers by use of three mirror model in the case of steady state. Further insights into the dynamic properties of diode laser are developed by deriving the well-known Lang-Kobayashi equations with inclusion of single reflection or multiple reflections, depending on the case of external feedback strengths. The solution of equations explain the variety of influences of external feedback on the properties of diode lasers. We also introduce different schemes for the implementation of optical feedback.

5.1.1 *General effects of external optical feedback on diode lasers*

It is well-known that external optical feedback strongly affects the properties of semiconductor lasers, the returned light into laser cavity causes variations in the lasing threshold, output power, linewidth, and laser spectrum [Olsson and Tang (1981)]. Most of the models dealing with these problems neglect the multiple reflections in the external cavity and simply incorporate the optical feedback by adding a time delayed feedback term to the standard laser equations [Lang and Kobayashi (1980) ; Hirota and Suematsu (1979)], some take into account multiple reflections [Osmundsen and Gade (1983); Hjelme and Mickelson (1987); Hui and Tao (1989); Zorabedian (1994)], which depend on the strength of the feedback. If the laser is antireflection (AR) coated ($R_2 \ll R_3$ as shown in Fig. 5.1), then only a single external-cavity round trip needs to be considered.

5.1.1.1 *Three-mirror laser cavity model*

The extended feedback model of the Fabry-Perot diode laser with external cavity is based on the three-mirror model [Petemann (1988)] as indicated in Fig. 5.1. M_1, M_2, and M_3 denote the two facets of the F-P laser and the external mirror, R_1 and R_2 are the real numbers of power reflectivities of mirrors M_1 and M_2, respectively, R_3 of external mirror. d and L are

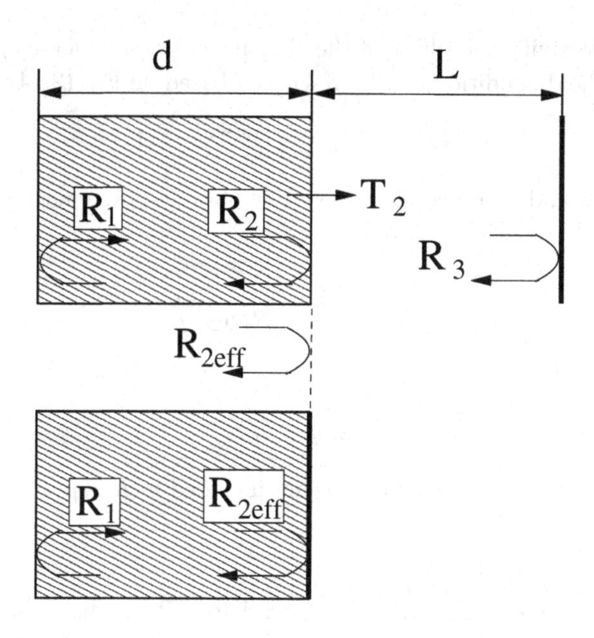

Fig. 5.1 External cavity laser and equivalent cavity with effective mirror to model the external section.

the lengths of the laser diode cavity and the external cavity, respectively. The reflection from the external mirror M_3 can be treated by combining it with the reflection of the laser end facet M_2, ending up with an effective reflectance $\Re_2$, also shown in Fig. 5.1 is an equivalent two-mirror cavity which replaces the passive section by an effective mirror with power reflectivity $\Re_2$. This substitution is valid for steady state analysis, but it will not necessarily proper model for dynamic operation. The power effective reflectivity is given by

$$\Re_2 = \sqrt{R_2} + \frac{1 - R_2}{\sqrt{R_2}} \sum_{k=1}^{\infty} [\sqrt{R_2 R_3} e^{-j\omega\tau_L}]^k, \tag{5.1}$$

it can be further simplified to

$$\Re_2 = \sqrt{R_2} + \frac{R_3(1 - R_2)e^{-j\phi_L}}{(1 + \sqrt{R_2 R_3})e^{-j\omega\tau_L}} = R_{2eff}e^{-j\phi_L}, \tag{5.2}$$

where $\phi_L = \omega\tau_L$, and ω is the angular frequency of the diode laser, τ_L is the round trip delay in the external cavity given by $\tau_L = 2L/c$, where c speed of light in the vacuum. R_{2eff} and ϕ_L are the notations representing the power reflectivity and phase of the effective reflectance $\Re_2$, respectively.

The threshold condition can be rewritten based on Eq. (2.14)

$$\sqrt{R_1 R_{2eff}} e^{2(\Gamma g-\alpha)d} e^{-j(\phi+\phi_L)} = 1. \tag{5.3}$$

Then, the gain and phase conditions can be obtained by replacing R_2 with $\Re_2$,

$$\Gamma g = \alpha + \frac{1}{2d} \ln \frac{1}{\sqrt{R_1 R_{2eff}}}, \tag{5.4}$$

and

$$\phi = 2\pi q - \phi_L, \tag{5.5}$$

any losses incurred in external mirror are included in R_{2eff}.

5.1.1.2 *External cavity modes*

In the absence of feedback, Eq. (5.5) for Fabry-Perot diode laser cavity becomes

$$\phi = 2\pi q. \tag{5.6}$$

We consider a single-mode laser characterized by the integer q_0 with a lasing wavelength $\lambda_0 = 2nd/q_0$. For small deviation in λ from λ_0, ϕ is a linearly decreasing function of λ, which is given by

$$\phi = 2\pi q_0 - 4\pi nd(\lambda - \lambda_0)/\lambda_0^2. \tag{5.7}$$

When feedback is present, the mode of the three-mirror cavity must satisfy Eq. (5.5) as well, but ϕ is no longer a linear function of λ. In general, the phase condition has multiple solutions for certain combination of the external cavity length L and the external reflection coefficient R_3. This is illustrated by plots of ϕ as a function of λ for increasing levels of the feedback in Fig. 5.2.

Generally, if R_3 is less than some value, then $\phi(\lambda)$ will decrease monotonically and the laser will remain single-mode regardless of the phase of the feedback. For large feedback values, $\phi(\lambda)$ is no longer monotonically decreasing and the laser will exhibit multiple external cavity modes for at least some range of phase of the feedback. Finally, for R_3 greater than a second value for which a minimum in $\phi(\lambda)$ is lower than the next two subsequent maxima, as indicated in Fig. 5.2(c), the laser will be multimode

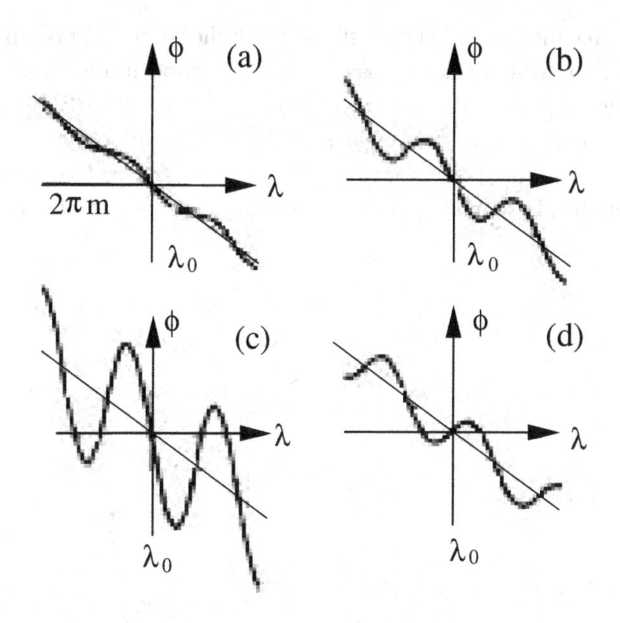

Fig. 5.2 Wavelength dependence of round-trip phase shift in external cavity laser plotted for different levels of feedback: (a) weak feedback, single-mode operation; (b) moderate feedback, single-mode operation; (c) strong feedback, multiple external cavity mode operation. In (d), the feedback level is same as (b), but the phase in the external cavity is shifted by π rads to give multimode operation. In (a)-(d) the thin line corresponds to no feedback ($R_3 = 0$). Adapted with permission from *IEEE J. Quantum Electron.* **QE-18**, 4, pp. 555-563. Goldberg et al. (1982).

regardless of the phase of the feedback. Thus, an external cavity or etalon can be used to filter out unwanted modes.

Figure 5.3 illustrates how the length of the external cavity determines the selection of the modes for the three cases of (a) L<d,(b) L$\sim$ d, and (c) L>d [Coldren and Corzine (1995)]. It presents the variations of mirror loss α_m, and the generic net gain curve $\Gamma g - \alpha$ versus wavelength along with indications of the mode locations. It is worth to note that the maxima in R_{2eff} corresponds to minima in α_m.

In Fig. 5.3(a), the external cavity is somewhat shorter than laser diode, the modes of the laser diode cavity will be more closely spaced than the minima in α_m. In this case a single loss minimum can effectively select a single axial mode of the active cavity if α_m varies enough. If the lengths of the laser diode and cavity are comparable as indicated in Fig. 5.3(b), the resonances of both cavities are spaced by about the same amount, and the active cavity modes will slowly slide across the minima of α_m providing an action similar to a vernier effects. Relatively good mode suppression is possible if the beat period is not too large or too small. In the third case of Fig. 5.3(c), good mode suppression is generally not possible unless the external cavity mirror itself is a filter. In fact, a grating mirror is widely used to select the single mode from diode lasers with a long external cavity.

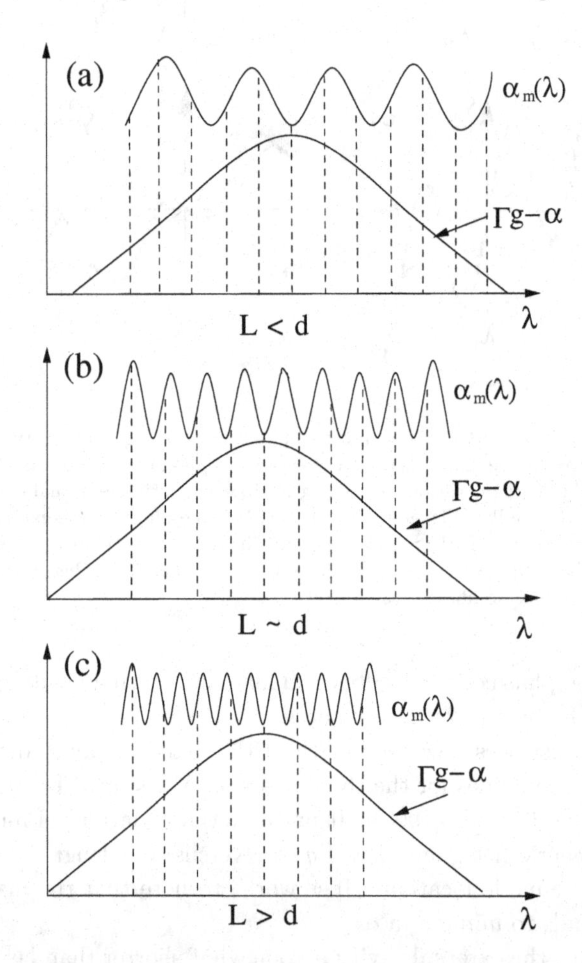

Fig. 5.3 Schematic illustration of net propagation gain $\Gamma g - \alpha$, and the net mirror loss α_m as a function of wavelength for external cavities, (a) $L < d$, (b) $L \sim d$, (c) $L > d$.

The influence of the optical feedback on diode laser properties with external cavity lengths ranging from $1 - 2$ cm has been investigated [Lang and Kobayashi (1980)]. Experiments have shown the periodic change in the output power, the laser spectrum change with the mirror distance, and DC excitation current, which arises from the change in interference conditions in the compound laser cavity. It has also been found that the external optical feedback can cause the laser to be multistage and to behave the hysteresis phenomena.

5.1.1.3 *Dynamical properties*

Some insights of the dynamical properties of semiconductor lasers can be gained from the characteristics of the light power versus current of diode laser when the diode laser is coupled to external cavities [Besnard *et. al* (1993)]. It was found that slight change of experimental conditions may result in considerable changes of the L-I curves, especially in the vicinity of the kink. Figure 5.4 shows five typical cases with L=30 cm, the characteristic curve without feedback is drawn for comparison and for quick estimation of the feedback levels. Each of the L-I curves (c) and (d) corresponds to a situation with optimum feedback, curves (a) and (b) to poor alignment. Curves (c) and (d) demonstrate the strong dependence of the system on the exact experimental conditions. For low injection currents, the two curves follow an identical path reminiscent of coherent feedback effects, they separate into two different branches above the solitary laser threshold. In the upper branch [curve (d)], the output intensity remains stable along the whole curve, while it comes out in the form of an enhanced noise beyond the transitions region c1-c2 of the curve (c).

L-I curves (b) and (c) have been obtained with poor alignment, their properties are summarized as follows: i) The compound cavity nature of the system still dominate even below the threshold, and fewer feedback into the active region, because of the lower Q of the external cavity as compared to the perfect alignment case; ii) When the injection current is increased with the exact external cavity geometry, the smooth kink in the vicinity of the solitary laser threshold is observed [curve (a)]. In some other cases, the kink does not show up as average output power is increased [curve (a′)].

The phenomena described above can be explained by the well-known Lang and Kobayashi [Lang and Kobayashi (1980)] rate-equation, that governs the dynamical properties of an external cavity laser,

$$\frac{dE(t)}{dt} = i\omega_0 + \frac{\Delta G}{2}(1 + i\kappa)E(t) + \xi E(t - \tau), \tag{5.8}$$

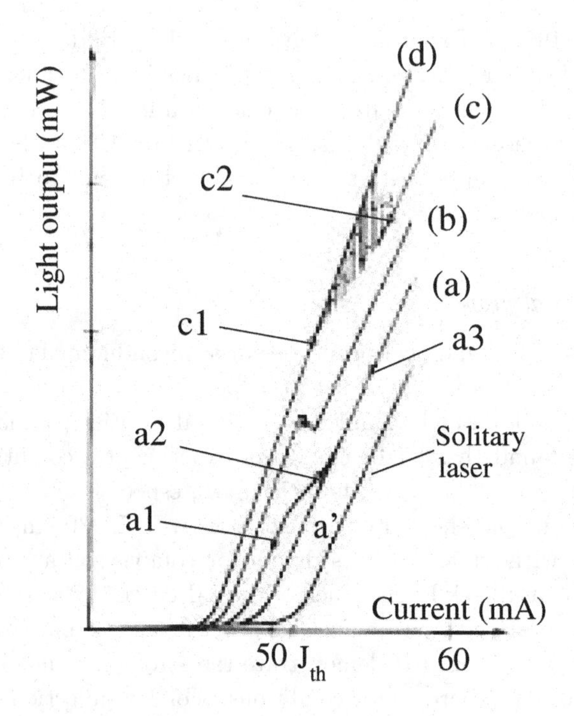

Fig. 5.4 Various light-output versus current characteristics obtained with slight variations in the optical feedback configuration. Adapted with permission from *IEEE J. Quantum Electron.* **29**, 5, pp. 1271-1284. Bernard and Duraffoug (1993).

and

$$\frac{dN(t)}{dt} = J - \frac{N(t)}{\tau} - G(N)|E(t)|^2, \tag{5.9}$$

in which G(N) is proportional to the difference between the actual gain and the gain at transparency:

$$G(N) = (N - N_0)\Gamma G_N. \tag{5.10}$$

The net gain

$$\Delta G(N) = G(N) - \frac{1}{\tau_p}, \tag{5.11}$$

depends on the rate of loss $1/\tau_p$ of the cavity, the coefficient of linear gain of the medium G_N and the actual carrier density N, Γ is the filling factor.

ξ is the feedback strength, that is given by

$$\xi = \frac{1 - R_3}{\tau_L}(R_2/R_3), \tag{5.12}$$

where τ_L is the round trip delayed time in the laser cavity. τ is the spontaneous carrier lifetime, J is the injection current, ω_0 is the lasing pulsation at solitary laser threshold in the absence of reflection, κ the phase amplitude coupling coefficient. Parameters used in these equations and their typical values are listed in Table 5.1.

Table 5.1 Parameters for expression of external cavity feedback effects.

Parameters	Expression	Typical values
LD cavity length	d	$300 \sim 500\ \mu m$
width of active zone	w	$\sim 5\mu m$
thickness of active zone	t	$\sim 0.1\ \mu m$
refractive index	n	3.5
reflectivity at the laser end facets	$R_1 = R_2 = (\frac{n-1}{n+1})^2$	0.31
roundtrip delay	$\tau_L = \frac{2L}{c}$	$\sim$ns
photon lifetime	τ_p	$\sim$2 ps
carrier lifetime	τ	ps$\sim$ns
confinement factor	Γ	0.25
carrier density at transparency	N_0	$\sim 1 \cdot 10^{24}/m^3$
carrier density at threshold	N_{th}	$\sim 2 \cdot 10^{24}/m^3$
differential gain	$G_N = (\frac{\partial G}{\partial N})_{N=N_0}$	$\sim 3 \cdot 10^{-12} m^3 s^{-1}$
gain compression efficient	ϵ^{NL}	$\sim 2.5 \cdot 10^{-17} m^3$

The field can be written as:

$$E(t) \sim \varepsilon(t) \exp[j(\omega t + \phi(t))], \tag{5.13}$$

and for optical intensity $I(t) = \varepsilon^2(t)$. Eqs. (5.8) and (5.9) take into account only a single feedback term, thus they are valid in the limitation of the case of weak feedback, i.e., $\xi\tau_L \ll 1$. The inclusion of multiple feedback contributions yields the improved rate equations:

$$\begin{aligned} \frac{dI(t)}{dt} &= g(N(t) - N_{th})I(t) - \epsilon^{NL} g(N(t) - N_0)I^2(t) \\ &\quad + \frac{2}{\tau_L} Re[\ln(f)]I(t) + \beta\frac{N(t)}{\tau} + F_l(t), \end{aligned} \tag{5.14}$$

$$\frac{d\phi(t)}{dt} = \omega_0 - \omega + \frac{\kappa}{2} g(N(t) - N_{th}) + \frac{1}{\tau_L} Im[\ln(f)]I(t) + F_\phi(t), \tag{5.15}$$

$$\frac{dN(t)}{dt} = J - \frac{N(t)}{\tau} - g(N(t) - N_0)I(t) + \epsilon^{NL} g(N(t) - N_0)I^2(t) + F_N(t), \tag{5.16}$$

which are valid for arbitrary levels of feedback. g is ΓG_N, F_N, F_l, F_ϕ are Langevin noise terms, $\epsilon^{NL} g(N(t) - N_0)I^2(t)$ describes the saturation effect. The coupling coefficient f is a sum over all the feedback terms and can be written as:

$$f = 1 + \sqrt{1 - R_2}\frac{\sqrt{R_3}}{R_2}\sum_{p=1}^{\infty}(-\sqrt{R_3 R_2})^{p-1} R_{eff}(p)\frac{\sqrt{I(t - p\tau)}}{I(t)} e^{j\theta^p(t) - p\psi}, \tag{5.17}$$

where $\psi = \omega\tau$ is the round trip phase change in the external cavity and $\theta^p(t) = \phi(t - p\tau) - \phi(t)$. $R_{eff}(p)$ is the effective reflectance for the p-th round trip beam.

This set of Eqs. (5.14)~ (5.16) have been the subject of the extensive studies on the behavior of strongly asymmetric external semiconductor lasers [Meziane *et. al.* (1995); Langley *et. al.* (1994); Langley *et. al.* (1995)]. Numerical solutions yield a noisy time trace, its correlation to the so-called coherence collapse has been proposed [Lenstra *et. al.* (1985); Dente *et. al.* (1988)]. Most of the various experimentally obtained phenomena seem to be contained in these equations owing to the delayed nature of the feedback term which renders the system of infinitely high dimension. Coherence collapse [Dente *et. al.* (1988); Schunk and Petermann (1988); Fischer *et. al.* (1996)], low-frequency intensity fluctuations [Mork *et. al.* (1988); Takiguchi *et. al.* (1998)], intermittent of noise [Sacher *et. al.* (1989)], as well as low-frequency intensity self oscillation with multiple delayed terms [Park *et. al.* (1990)], have been studied following extensive numerical calculations based on the various values of the system's control parameters.

5.1.2 *Implementation of optical feedback*

In general, the uncontrolled optical feedback in the design of external cavity diode laser systems is undesirable. There are a couple of ways to implement the optical feedback as follows [Fox *et. al.* (1997)]:

i) Using simple optical elements, such as mirrors,[Salathe (1979); Akerman *et. al.* (1971); Voumard *et. al.* (1977); Harrison and Mooradian (1989)] to feed some of its output power back to the laser diode, the diode's facet and the external optical elements constitute a net resonator as shown in Fig 5.1. Optical feedback-induced changes in the output spectra of several GaAlAs lasers operation at 830 nm were described [Goldberg *et. al.* (1982)]. The feedback radiation obtained from a mirror 60 cm away from the laser is controlled in its intensity and phase. Spectral line narrowing or broadening is observed in each laser depending on the feedback conditions. Minimum linewidth observed with feedback is less than 100 kHz. Improved wavelength stability is also obtained with optical feedback resulting in 15 dB less phase noise. An analytical model for the three-mirror cavity has been developed to explain these observations. The influence of optical feedback on the laser frequency spectrum and on the threshold gain, taking into account multiple reflections, have been analyzed [Osmundsen and Gade (1983)]. The first systematic study of the effects of feedback asymmetry, as determined by misalignment of an external mirror, such as tilted mirror, on the characteristics of external-cavity semiconductor lasers has been demonstrated [Seo *et. al.* (1988)].

ii) Applying weak optical coupling of the laser's output to a high-Q optical resonator [Dahmani *et. al.* (1987); Hollberg and Ohtsu (1988); Li and Telle (1989)]. With the appropriate optical geometry, the laser optically locks itself to the resonance of a separate Fabry-Perot reference cavity, the method depends on the occurrence of optical feedback only at the resonance of a high-Q reference cavity, and is used to stabilize laser frequencies, and thus to reduce linewidths by a factor of 1000 from 20 MHz to approximately 20 kHz. A confocal Fabry-Perot (CFP) cavity [Laurent *et. al.* (1989); Hemmerich *et. al.* (1990a)] is used to feedback the beam from the diode laser and provide resonant optical stabilization of the semiconductor laser at same time[Hollberg and Ohtsu (1988)]. The choice of a confocal cavity greatly facilitates the mode coupling of the diode laser to the CFP, the CFP is tilted in such a way that it can be considered a four-ports device which eliminates the reflection-like beam, but still provides optical feedback to the laser from the transmission-like beam. It is demonstrated experimentally that static frequency noise reduction of 50$\sim$60 dB is achieved and a dramatic reduction of the laser linewidth from 20 MHz to less than 4 kHz is obtained.

iii) Using the antireflection (AR) coating on the diode laser chip and some external optics to provide the laser resonator, the external optics may contain frequency selective elements such as grating and/or etalon [Fleming and Mooradian (1981); Wyatt and Devlin (1983); de Labachelerie and Cerez (1985); Mittelstein *et. al.* (1989); Harvey and Myatt (1991); Boshier *et. al.*

(1991); Schremer and Tang (1990)]. For grating-tuned feedback with AR coated diode laser ($R_2 \ll R_3$), Eq. (5.2) can be simplified to single round-trip reflection, which is given by

$$\Re_2 = \sqrt{R_2} + (1 - R_2)\sqrt{R_3(\omega)}e^{-j2\omega L/c}, \tag{5.18}$$

where ω is the angular frequency of the diode laser, $R_3(\omega)$ is the round trip power transmission of the single-sided external cavity into the lasing mode just outside the diode laser, it has peak at the Littrow frequency ω_0, i.e., the frequency which is retroreflected by the grating. As ω deviates from ω_0, the coupling back from the grating into laser mode just outside the facet becomes progressively smaller. Therefore, the spectra reflectivity $\Re_2$ is changed from R_2 alone only near the Littrow frequency. The threshold condition after one round trip in the diode laser resonator can then be written as

$$\sqrt{R_1 R_{2eff}}e^{2[g(\omega)-\alpha]d}e^{-j2n(\omega)\omega d/c} = 1. \tag{5.19}$$

where $g(\omega)$ is the modal gain of the semiconductor medium, α is its distributed modal loss, and d is its length.

The requirements for the modal gain at threshold and the compound cavity oscillation frequencies then become

$$g(\omega) = \alpha + \frac{1}{2d}\ln\frac{1}{\sqrt{R_1 R_{2eff}}}, \tag{5.20}$$

and

$$2\pi q = 2n(\omega)\omega d/c + \Phi, \tag{5.21}$$

where

$$\Phi = Arg(R_{2eff}) = \arctan[\frac{(1 - R_2)\sqrt{R_3(\omega)}\sin(2\omega L/c)}{\sqrt{R_2} + (1 - R_2)\sqrt{R_3(\omega)}\cos(2\omega L/c)}], \tag{5.22}$$

and q is an integer. The phase condition indicated by Eq. (5.21) is satisfied for a group of frequencies (as discussed previous section) clustered near the grating-selected longitudinal mode of solitary diode laser, the spacing within the group is approximately $\pi L/c$. The corresponding threshold of these modes varies. In general, there will be one mode which corresponds closely to constructive interference from external cavity if the number of solutions of Eq. (5.21) in the vicinity of each mode of solitary diode laser is sufficiently large.

In the case of constructive interference at $\omega = \omega_0$, the gain requirements is minimized, and Eq. (5.20) becomes

$$g(\omega_0) = \alpha + \frac{1}{2d}\ln[\frac{1}{\sqrt{R_1R_2} + (1-R_2)|\sqrt{R_3(\omega)}|}]. \quad (5.23)$$

If the round trip gain first reach unity at the tuned frequency ω_0, the compound diode laser will lase at the corresponding photon energy. The tuning range is then determined by the flatness of the gain spectrum relative to the amount of grating feedback. If the change in gain is small over a wide spectral range, then broadband tuning is possible even with a small amount of external feedback.

(iv) Combining two different diode laser concepts: the diode laser with (a) feedback from a grating and (b) resonant optical feedback from a separate cavity. The novel concept is to unite the excellent tunability and well known reliability of grating diode lasers with resonant optical feedback. The experiment is based on an AR coated diode laser emitting at the wavelength of 852 nm. An overall tuning range of 36.4 nm, a continuous tuning range of 45.1 GHz, and a narrow linewidth below 60 kHz have been obtained [Wicht *et. al.* (2004)].

5.2 Spectral characteristics of ECDLs

In chapter 2, we have introduced the spectral characteristics of solitary diode laser. In general, the wavelength of the solitary laser emission cannot be completely controlled, a typical device may operate in several modes, with mode competition and hopping. Moreover, variation of the injection current in an isolated laser diode causes simultaneous and inseparable shifts in the frequency, the output power, and the allocation of power among the different modes for a multimode laser. As described in the foregoing sections, the external-cavity laser with feedback operates in a single mode with linewidth considerably less than that of a solitary laser diode. The insertion of dispersive element in the external-cavity configuration allows one to select and tune the emission wavelength by external control, without the complication associated with the variation of the pumping level.

In the following, we are concerned with the primary characteristics of external cavity diode laser such as output power, single-mode tunability, linewidth, and frequency dependence on the diode laser temperature.

5.2.0.1 *Output power*

The laser configuration is schematically shown in Fig. 5.5. For maximum grating dispersion, the diode laser should be mounted with its junction plane parallel to the grating rulings. The left lens collects the diode radiation and collimates it onto the diffraction grating, which is mounted in the Littrow configuration, whereby the first-order diffracted beam is reflected collinear with the incident beam and re-imaged on the left diode facet. The radiation from the right facet is collimated onto the output coupler, a dielectric-coated, partially reflecting plane mirror.

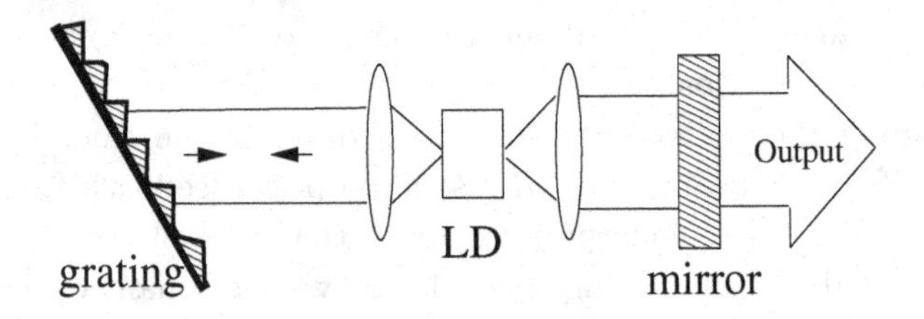

Fig. 5.5 Schematic diagram of two-sided external cavity diode laser. Adapted with permission from *IEEE J. Quantum Electron.* **QE-17**, 1, pp. 44-59. Fleming and Mooradian (1981).

The laser output power is plotted against the injection current with solitary laser and ECDLs system. The stimulated power leaving the cavity through the output coupler can be obtained based on Eq. (2.25),

$$P_{out} = \eta_i \frac{h\nu}{e}(I - I_{th})\left(\frac{\ln \frac{1}{\sqrt{R_3}}}{\alpha l + \ln \frac{1}{\sqrt{T_l^4 R_g R_3}}}\right), \tag{5.24}$$

where η_i is the internal radiative recombination efficiency, R_g and R_3 are the grating and output coupler reflectivities, respectively. T_l is the lens transmission. One yields the threshold conditions

$$\Gamma g = \alpha + \frac{1}{l} \ln \frac{1}{\sqrt{T^4 R_3 R_g}} \tag{5.25}$$

where Γ is the mode confinement factor and g is threshold gain. The above equations for power output and threshold condition indicate that, for a given injection current, the external-cavity laser generally has a somewhat lower power output than that of the solitary diode laser, (despite the fact that the differential efficiency may be nearly same), moreover, has higher threshold current, as shown in Fig. 5.6

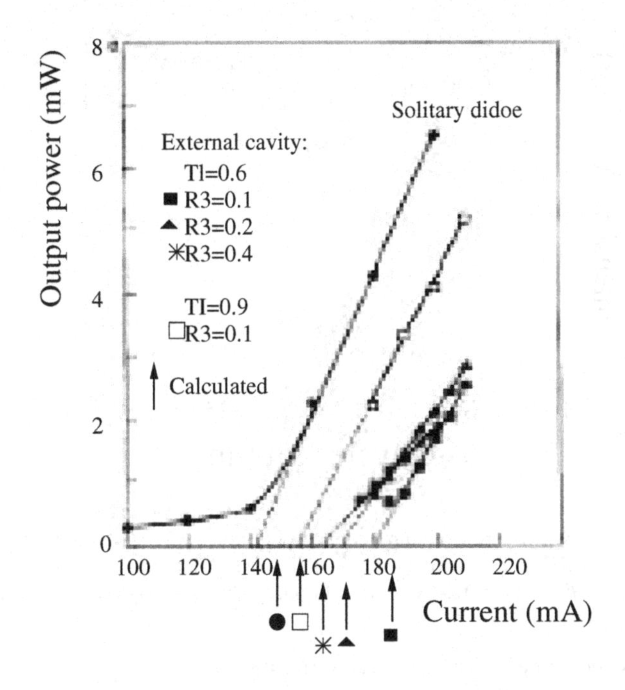

Fig. 5.6 Optical output power of the solitary diode compared with that of the external-cavity laser with various mirror and lens combinations. For solitary diode, $R_g = 0, R_1 = R_2 = 0.32$, for external cavity, $R_g = 0.9, R_1 = R_2 = 0.01$. Adapted with permission from *IEEE J. Quantum Electron.* **QE-17**, 1, pp. 44-59. Fleming and Mooradian (1981).

5.2.0.2 *Single-mode tunability*

The output of the solitary diode is usually multimode, as exemplified in Fig. 5.7. The external cavity diode laser operates in a single frequency, typified by the spectrum in the inset of Fig. 5.7. Fig. 5.8 shows the typical laser mode tuning range, which extends over more than 10 nm for an output coupler reflectivity of 20%. Continuous tunability of the laser emission depends on the precision with which the cavity aligned and upon the quality of the antireflection coated facet of diode laser. The laser mode frequency could be adjusted to any desired point between adjacent mode frequencies of the solitary diode. This fact indicates that the single-mode behavior of the external-cavity laser is not critically dependent on the frequency selectivity of the intracavity Fabry-Perot etalon composed of the diode facet, but is rather a result of the essentially homogeneous spectral saturation of the laser gain. Neither is the single-mode behavior simply a result of the

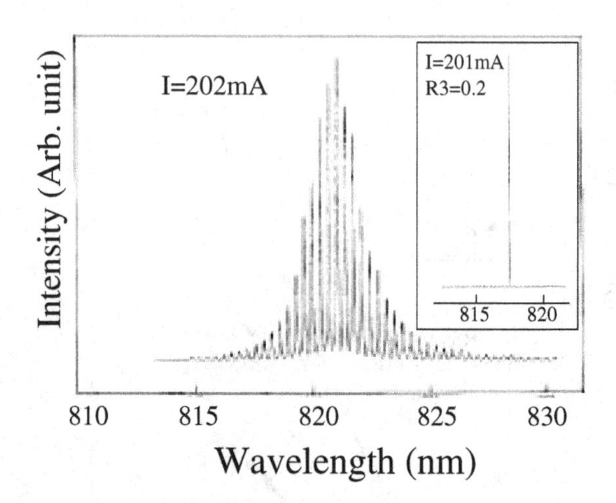

Fig. 5.7 Spectrum of solitary laser diode before operation in the external cavity. Inset: spectrum of the diode operated in the external cavity. Adapted with permission from *IEEE J. Quantum Electron.* **QE-17**, 1, pp. 44-59. Fleming and Mooradian (1981).

wavelength selectivity provided by the grating.

5.2.0.3 *Linewidth*

The most striking feature of the stable external-cavity laser is its narrow linewidth. By extending the optical cavity length, the spontaneous-recombination phase fluctuation in the laser linewidth can be dramatically reduced. The power spectrum of the electric field is Lorentzian, with a full width at half maximum (FWHM) given by the modified Schawlow-Townes formula, [Welford and Mooradian (1982); Henry (1986)]

$$\Delta\nu_q = \frac{h\nu_q g n_{sp} (\Delta\nu_g)^2}{P_o} \alpha_t \cdot (1+\beta^2), \tag{5.26}$$

where P_o is the power in the mode, n_{sp} is the number of spontaneous emission photons in the mode, g is the gain, h is Planck constant, ν_q is frequency of laser. The total loss $\alpha_t = \alpha - \ln\sqrt{R_1 R_2}$ for solitary diode; and $\alpha_t = \alpha - \ln\sqrt{T^4 R_1 R_2}$ for external cavity diode laser, R_1 and R_2 are the facet reflectivities, respectively. The spectral linewidth enhancement factor β is given by $(dn/dG)(dG/DN)$, the typical values range from 2 to 8 [Osinski (1987)]. $\Delta\nu_g$ is the bandwidth (FWHM) of the Fabry-Perot cavity, it is related to the photon lifetime τ_p and consequently, to the single-pass

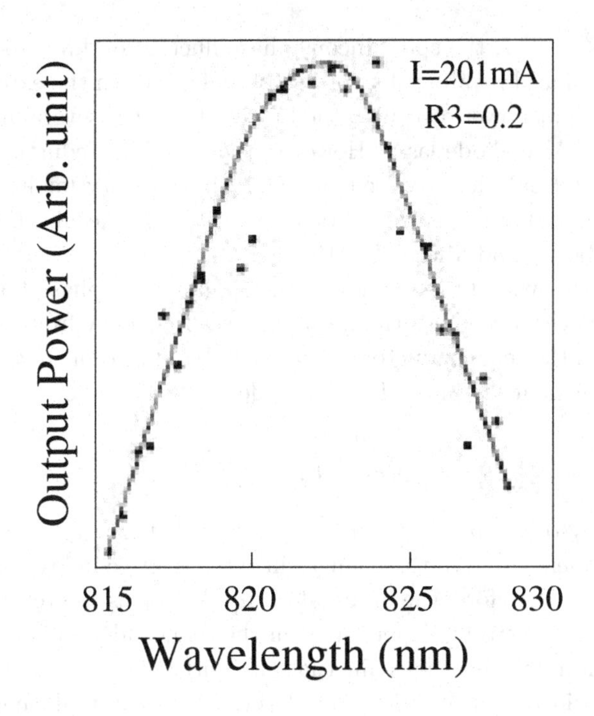

Fig. 5.8 Stimulated power tuning spectrum of the external-cavity laser. Adapted with permission from *IEEE J. Quantum Electron.* **QE-17**, 1, pp. 44-59. Fleming and Mooradian (1981).

cavity loss. For a solitary diode laser, the appropriate expression for the cavity bandwidth is

$$\Delta\nu_{gSD} = \frac{1}{2\pi\tau_p} = \frac{c}{2\pi nd}, \tag{5.27}$$

where n is the refractive index of the medium. Eq. (5.27) must be modified when the diode is operated in an external cavity, where the photon lifetime is significantly longer due to loss-free propagation over a distance $L \gg nd$:

$$\Delta\nu_{gEC} = \frac{c}{2\pi(nd + L)}. \tag{5.28}$$

In terms of linewidth for no feedback $\Delta\nu_{gSD}$, and for a given mode power P_o, we have

$$\Delta\nu_{gEC} = \frac{\Delta\nu_{gSD}}{(1+\gamma)}, \tag{5.29}$$

where $\gamma = L/nd \gg 1$, the spontaneous phase fluctuation limit to the laser linewidth is reduced by about five orders of magnitude in the external cavity. For an output power of 1 mW, the linewidth limit is several megahertz in a typical GaAlAs diode laser. However, the linewidth reduction proportionality is valid only for low feedback (<1%). A theory for linewidth of steady state external cavity valid for arbitrary strong feedback has been considered [Hjelme and Mickelson (1987)].

Concomitant with its reduction of the spontaneous phase fluctuation, the external cavity decouples the resonant laser frequency from the strong dependence on the semiconductor refractive index. Since the optical length of the external cavity is (nd + L), the mode frequencies are

$$\nu_q = \frac{qc}{2(nd + L)}, q = 1, 2, ... \tag{5.30}$$

where $\nu_q^{-1}(\partial\nu_q/\partial_n) = n^{-1}[(1+\gamma)]^{-1}$. For $\gamma \gg 1$, the relative changes in the mode frequency due to the changes in the refractive index are reduced by the factor γ, therefore, the external cavity decouples the resonant laser frequency from the strong dependence on the semiconductor refractive index, while the refractive index fluctuation contribute significantly to the observed linewidth of a solitary diode laser, for an external-cavity lasers, they are negligible.

5.2.0.4 *Wavelength dependence of temperature*

The temperature sensitivity of laser threshold current in single mode wavelength tunable diode laser has been measured in the temperature range 293 K (20 °C) $\leq$ T $\leq$ 355 K (82 °C) and the wavelength range 1.23 $\mu m \leq \lambda \leq 1.35$ μm [Ogorman and Levi (1993)]. When proper account is taken of peak gain variation with temperature, the laser threshold current dependence on temperature is insensitive to lasing wavelength over a wide tuning range. The variation of lasing threshold with temperature can be fitted to the phenomenological expression

$$I_{th} = I_0 e^{T/T_0}. \tag{5.31}$$

where T_0 is the overall characteristic temperature, note that small values of T_0 indicate a larger dependence on temperature. The devices used in the experiment are a standard bulk active InGaAsP buried heterostructure design. The experimental arrangement for the external cavity is shown in Fig. 5.9, light output from a high quality antireflection coated facet is

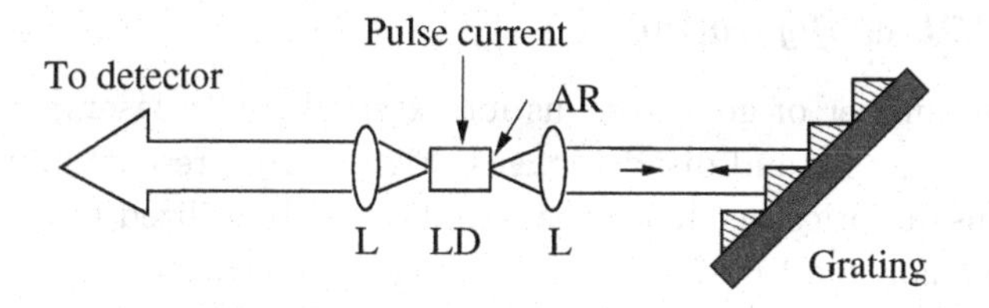

Fig. 5.9 Schematic diagram of external cavity laser diode laser. LD: laser diode, L: lens, AR: antireflection coating.

efficiently coupled to the external cavity by a low loss lens. The external cavity of length 20 cm is closed by a 600 groove/mm diffraction grating, that allows laser emission to be tuned in a single instrument limited line across the semiconductor gain spectrum.

The experimental results show that over a large temperature range (293 K$< T <$355 K) and wide tuning range 1.23 $\mu m \leq \lambda \leq 1.35$ μm the threshold current can be well characterized by a simple expression:

$$I_{th}(\lambda, T) = I_0(\lambda)e^{T/T_0}, \tag{5.32}$$

where $\lambda = \lambda - \xi T$. Eq. (5.32) is a remarkable result. Since the diffraction grating is blazed at $\lambda = 1.25$ μm and the optics coupling to the cavity has broadband antireflection coating, the magnitude of cavity coupling efficiency does not vary substantially over the range of detunings investigated. Consequently, the increased threshold current required with detuning from the gain peak is not a consequence of changing loss level, but of changing energy distribution of charges carriers in a forward biased laser diode.

5.3 System of tunable external cavity diode lasers

It can be seen from the previous sections that the optical feedback can accomplish the tuning of diode laser in the broad ranges and the narrowing of linewidth by use of some forms of optical elements. In this section, we utilize the dispersive high optical feedback power to obtain a large tuning range and narrow linewidth in different external cavity configurations. The optimal external cavity designs and alignments are examined. Diffraction gratings are the most commonly used dispersive feedback element, though intracavity etalon, prisms, electronically tuning birefringent filters, and acousto-optic tunable filter have also been widely used successfully, which will be discussed in the next chapter.

5.3.1 *ECDL configuration*

There are a number of good designs for external cavity lasers, they have their own advantages and disadvantages. Figure 5.10 presents four various constructions of optical feedback that can be mostly utilized to control the diode lasers. For good performance, antireflection coatings on the output facet of diode chip are highly desirable[Fox *et. al.* (1997)].

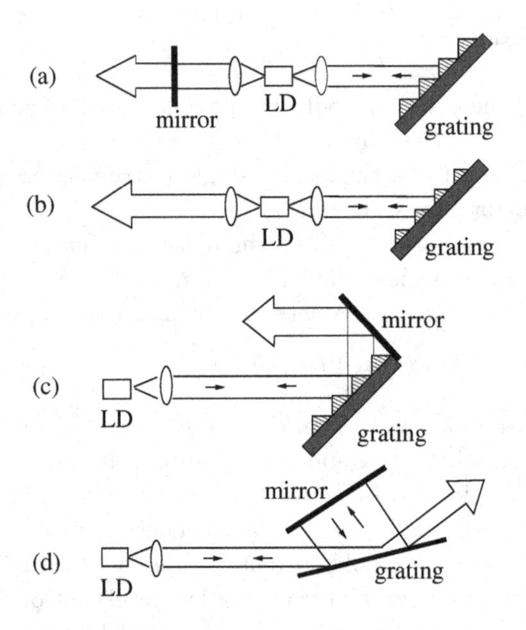

Fig. 5.10 Different optical feedback schemes: (a) and (b) a two-side external-cavity design, (c) a one-side external cavity designs in the Littrow configuration, (d) an external-cavity design using a grating in a grazing-incidence Littman-Metcalf scheme.

In Fig. 5.10(a), two-sided external cavity design is shown [Fleming and Mooradian (1981)], this scheme contains an diode chip with AR coated on both facets. Each extended-cavity section retroreflects into its respective facet. One of the extended-cavity sections contains the grating for wavelength selectivity, the other has only coupling output mirror. The configuration in Fig. 5.10(b) is similar to that in Fig. 5.10(a) by replacing the output mirror with optical filter. The two-sided external cavity as shown in Fig. 5.10(a) and (b) offer advantages of the increase of suppression of diode cavity resonance obtained by reducing both facet reflectivity and no movement of output beam when the grating is rotating to tune the wavelength.

On the other hand, it has disadvantages of special laser package to get the unobstructed access to both facets, and the increase of alignment difficulty and additional coupling loss associated with the second extended-cavity section.

Fig. 5.10(c) is the single-sided extended-cavity design in Littrow configuration, which is the most commonly used configuration for the following reasons: i) only one AR coated facet; ii) commercially available package with one facet output; iii) simplicity and easy alignment; iv) excellent performance. However, even with high quality facet coating, the effects of the residual diode cavity resonance are still observable and sometimes the cause of non-ideal behavior. There is also the disadvantage that the beam moves as the laser wavelength is tuned. A remedy to this problem is to use a mirror mounted with its surface perpendicular to the grating surface so as to form a retroreflector [Hawthorn *et. al.* (2001); Turner *et. al.* (2002)]. Rotation of both the grating and mirror together leaves the output beam direction unchanged, although there will be some beam displacement if the axis of rotation is not defined by the intersection of the grating and the mirror.

An extremely simple laser of this type has been developed [Arnold *et. al.* (1998); Lancaster *et. al.* (2000); Turner *et. al.* (2002)] by building from inexpensive commercial components with only a minor modifications. A 780 nm laser built to this design has an output power of 80 mW, a linewidth of 350 kHz. As shown in Fig. 5.11, the ECDL systems are based on the Littrow configuration design. Each consists of a Sanyo DL-7140-201 laser diode and aspheric collimating lens mounted in a collimation tube fixed to a modified mirror mount. A 10 kΩ thermistor sensor and Peltier thermoelectric cooler are used to stabilize the diode temperature. A gold-coated diffraction grating with 1800 lines/mm on a $15 \times 15 \times 3\ mm^3$ substrate provides wavelength-selective feedback with a typical diffraction efficiency of about 15 % up to 80 % of the intra-cavity power, which is directly reflected to form the output beam. The grating is attached to the front face of mirror mount, which provides vertical and horizontal grating adjustment. A 1 mm thick PZT piezoelectric transducer disk under the grating is used to modify the cavity length for fine frequency tuning. The output beam is reflected from a mirror attached to the grating arm. The double reflection from grating and mirror maintains a fixed output beam direction as the grating angle and lasing wavelength is adjusted. The output power is typically 40 mW at 780 nm, and the wavelength can be tuned discontinuously over a 10 nm range by rotation of the grating alone, and over a wider range with suitable temperature adjustment.

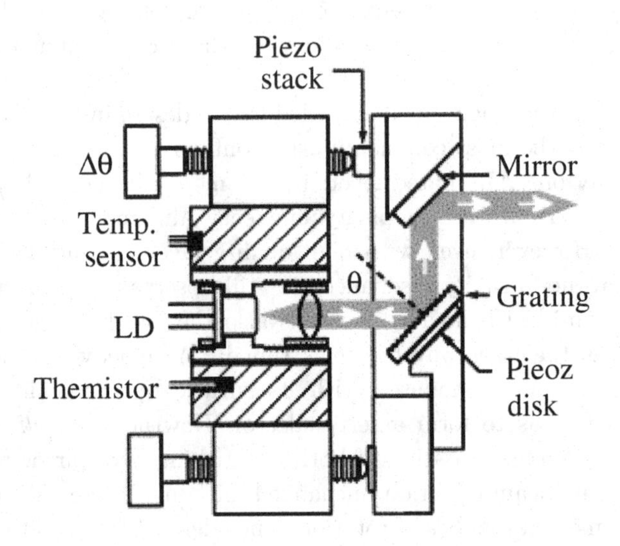

Fig. 5.11 External cavity diode laser design. An AlGaAs laser diode, aspheric collimation lens, and diffraction grating in a Littrow configuration, are mounted in a kinematic mirror mount. A thermistor temperature sensor is used for feedback to stabilize the laser temperature with a thermoelectric cooler (not shown). An LM35 semiconductor sensor provides a separate temperature readout. The mirror maintains a fixed output beam direction as the grating angle is adjusted. A piezoelectric stack is used to adjust the grating angle and hence wavelength (20 GHz/100 V) and a piece disk adjusts the cavity length for frequency locking feedback. Adapted with permission from *Opt. Commun.* **201**, pp. 394. Tuner et al. (2002).

The lasers also have a stacked piezoelectric transducer which drives the grating-mirror pivot arm, this stack alters the grating angle and the cavity length, allowing electronic wavelength adjustment of 20 GHz over the 100 V range of the stack. This allows greater scanning range and much safer voltages compared to the original design. Each laser is mounted to a heavy metal base to provide inertial and thermal damping. The base is isolated from the optical bench with viscoelastic polymer at the corners and enclosed with an aluminum cover, which is also isolated from the laser by strips of Sorbothane. The laser and saturated absorption optics are covered with an acrylic enclosure, the aluminum cover and acrylic enclosure shield the lasers from air currents, improve temperature stability, and suppress acoustic vibrations.

An external cavity diode laser system with diffraction-grating feedback has been developed near the cesium D1 transition at 894 nm, producing over 20 mW of single-mode power with a continuous tuning range of up to 25 GHz [Andalkar *et. al.* (2000)]. Compared to the usual Littrow configuration described above, their design consists of a pair of custom-built kinematic mounts, one holding the diode and collimating lens, and the other the diffraction grating. This allows one to easily and precisely optimize the laser output and diode temperature can be controlled with minimum effect on the cavity length.

Fig. 5.10(d) is an extended-cavity laser that uses a grating in a grazing-incidence configuration. This design, implemented with diode laser by Harvey et al [Harvey and Myatt (1991); Day *et. al.* (1995)], is often called the Littman-Metcalf configuration because they introduced the use of grazing-incidence grating to control the Dye lasers [Littman and Metcalf (1978); Littman (1984)].

It is a three-mirror cavity that consists of a high-reflection-coated rear facet of the diode laser, the lasing medium of the diode laser, the antireflection-coated front facet of the diode laser, a collimating lens, a diffraction gratin at grazing incident, and an external mirror. The zeroth-order reflection from the grating is the output of the laser. The first-order reflection from the grating is reflected back into the laser by the external mirror. One end of the laser cavity is the rear facet of the diode laser, and the other end is the external mirror. The arrangement is similar to the Littman [Littman and Metcalf (1978)] configuration for the pulse lasers. However, it differs in that here a lens is added to collimate the beam, the gain medium is in a waveguide, and the laser is continuous wave. This configuration has important advantages of significantly higher spectral resolving power and no movement of the output beam when the laser is tuned. A spectral width based on this configuration is determined by

$$\frac{\Delta\lambda}{\lambda} = \frac{\lambda}{\pi w \sin\theta}, \tag{5.33}$$

where $\Delta\lambda$ is the half width of the spectral distribution of the output laser at wavelength λ, w is the width of the illuminated part of the grating, and θ is the angle between the grating normal and incident beam. To reduce the spectral linewidth of the diode laser one tries to cover the laser beam to the grating by increasing the incident angle.

Fig. 5.12 (a) is a resonant optical-locking configuration using a confocal Fabry-Perot cavity [Dahmani *et. al.* (1987); Olsson *et. al.* (1987)]. In this system one uses the weak optical coupling of the laser output to a high-

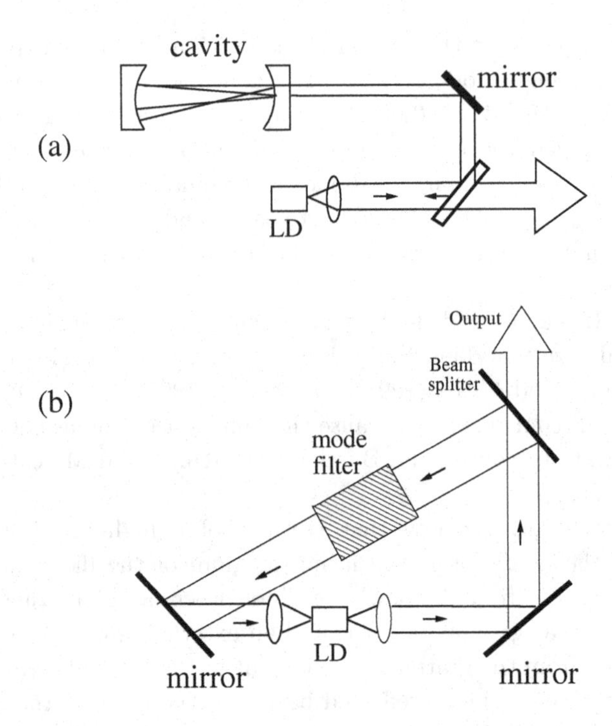

Fig. 5.12 Different optical feedback schemes: (a) a resonant optical-locking configuration using a confocal fabry-Perot cavity. (b) Ring cavity scheme.

Q optical resonator, the laser sees optical feedback from the Fabry-Perot cavity only when the laser's frequency matches a resonance of this cavity. In this way the laser's linewidth can be reduced to a few kilohertz and the laser's center frequency is stabilized to the cavity resonance. The laser is tuning by changing the length of Invar Fabry-Perot cavity with a PZT. One of the limitations of this system is that it requires some additional slow electronics to keep the laser locked to the same cavity diode for long time and keep the laser synchronized with the cavity mode for long scans. The major advantages of this system are that the linewidth is very narrow and the stability is determined by the external cavity. The disadvantages are the sensitivity to cavity laser separation and the fact that laser tuning range is still essentially the same that of the basic unstabilized laser since it is restricted to the weak-feedback regime.

Fig. 5.12(b) shows a ring-cavity laser, which is the most difficult type of the external cavity to align. Like the two-side external cavity, the ring-cavity has the advantage of the increase of solitary resonance suppression because of reflectance suppression on both facets. It can be made unidirectional by inserting an optical isolator into the cavity. An all fibers, widely tunable, single frequency, semiconductor ring laser has been constructed with a narrow linewidth of 350 kHz, tuning over 50 nm was electronically achieved by use of a fiber Fabry Perot filter [Chawki *et. al.* (1993)].

In contrast with Littman configuration described previously, the tuning mirror can be replaced with a second grating at Littrow angle [Wandt *et. al.* (1997)]. This double grating arrangement was first implemented theoretically by Littman [Littman (1978a)] and experimentally by [Shoshan and Oppenheim (1978)], respectively, in order to develop the narrow band pulsed dye lasers. As shown in Fig. 5.13, the passive bandwidth of the combination of the grazing-incidence grating and a Littrow grating is

$$\delta\nu_{dg} = \frac{2c}{\pi w\lambda(2\frac{q_1}{d_1\cos\beta_1} + \frac{q_2}{d_2\cos\beta_2})\frac{\cos\beta_1}{\cos\alpha_1}}, \tag{5.34}$$

where c is the velocity of light, w is the beam radius, d_1 and d_2 are the groove spacing of the grazing incidence grating and the Littrow grating, respectively, q_1 and q_2 the corresponding diffraction orders. α_1 is the angle of incidence, β_1 is the diffraction angle of the first grating, β_2 the Littrow angle of the second grating, and λ is the laser wavelength. Eq. (5.34) simplifies for the grating-mirror combination ($q_2 = 0$) to

$$\delta\nu_{sg} = \frac{cd_1}{\pi q_1 w\lambda}\cos\alpha_1. \tag{5.35}$$

By comparing Eqs. (5.34) and (5.35), one finds that the double-grating configuration permits a bandwidth that is a factor of $(1 + d_1\cos\beta_1/2d_2\cos\beta_2)$ smaller than that of the grating-mirror cavity ($q_1 = q_2 = 1$). At a fixed groove spacing of the grazing-incidence grating, this reduction is strongly dependent on the groove spacing of the Littrow grating. Using a Littrow grating with a grating period of 1200 g/mm and a typical angle of incidence of 80 ° at a wavelength of 770 nm, one obtained a factor of 1.3 of bandwidth reduction.

However, by using the same parameters but a grating with twice the above-mentioned grating period, one can achieve a reduction factor of 2.6. Therefore the angle of incidence α_1 can be reduced for operation with a given bandwidth. This reduces cavity losses, and hence increases the tuning range. In addition, the double grating arrangement allows one to synchronize the cavity mode scan and the feedback wavelength scan by a simple mechanical construction, and therefore wavelength tuning with mode

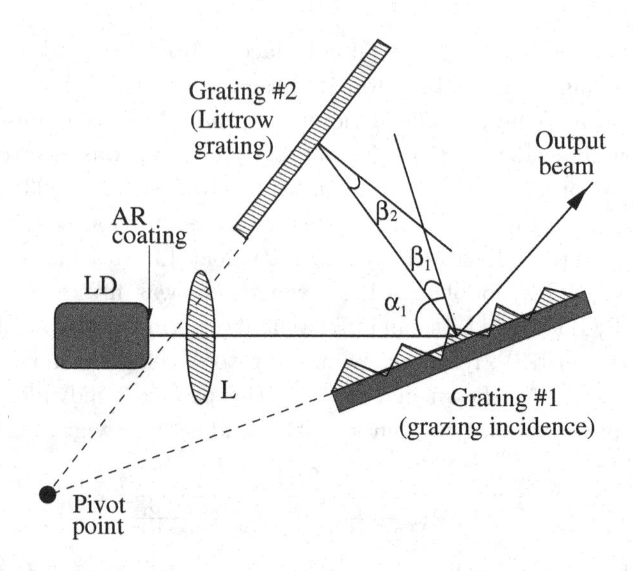

Fig. 5.13 Schematic of the double-grating arrangement. LD: laser diode, AR: antireflection, L: Collimator. Adapted with permission from *Opt. Lett.* **22**, 6, pp. 390-392. Wandt et al. (1997).

hop-free [Mcnicholl and Metcalf (1985)]. Finally, the dependence of the wavelength on the rotation angle of the tuning element is smaller in the double-grating design than that in grating mirror configuration. Hence the laser can be tuned with high-frequency resolution.

The tuning curve of an optimized 80 nm laser system is shown in Fig. 5.14(a), the laser is continuously tuned in a single longitudinal mode without mod hops in the wavelength from 805 nm to 840 nm at current 90 mA and a temperature 20 °C. At the center of the tuning curve at 820 nm, a maximum output power of 6 mW is achieved. The modulation of the output power arises from the weak residual reflection of the front facet of the laser diode, forming a low-quality resonator. The dependence of the output power on the wavelength for 775 nm external-cavity laser system is shown in Fig. 5.14(b). The laser is continuously tunable from 758 nm to 785 nm with an output power of 4.5 mW at the peak of the tuning curve at 775 nm. By using the same components in a single-grating Littman configuration but with an angle of incidence on the grating of 84 °, a continuously tuning range of 20 nm has been achieved.

The integration of micromachined mirror and laser diode to form miniaturized external-cavity tunable laser has been proposed [Kiang *et. al.*

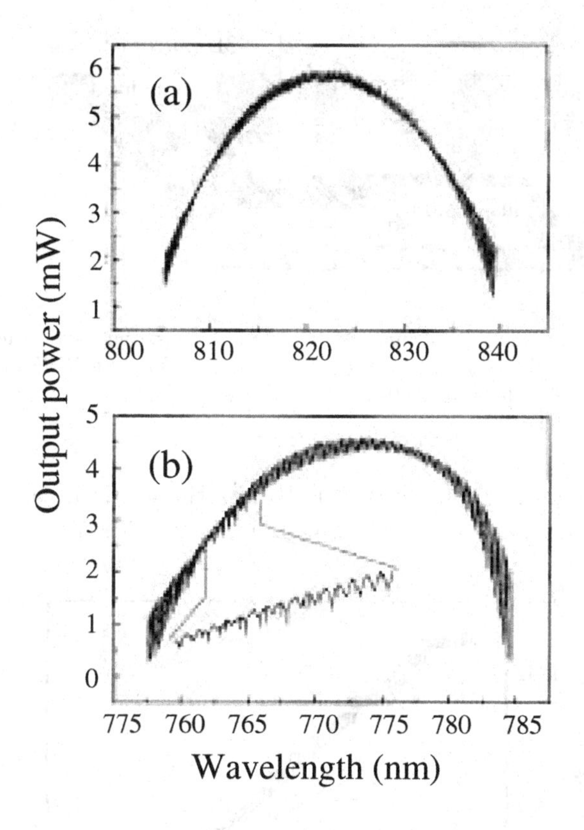

Fig. 5.14 Output power versus wavelength for the (a) 820 nm laser system and for the (b) 775 nm external-cavity laser. Wavelength tuning is performed without any mode hop across the entire spectra. Adapted with permission from *Opt. Lett.* **22**, 6, pp. 390-392. Wandt et al. (1997).

(1996)]. A micromachined tunable diode laser has been made by the integration of a surface micromachined 3-D mirror, and FP diode laser and butt-coupling optical fiber, as is shown in Fig. 5.15. The optical fiber is aligned very near to the other exit window of the laser diode with the intention of directly coupling the output laser beam from the laser diode without the need for optical coupling lenses. The 3-D mirror sitting on the a translating stage can be driven to move by the comb drive. The suspension beam are used to maintain the translating stage hanging over the substrate. The micromirror surface is located parallel to one of the exit window of the laser diode and reflect part of the laser beam back into the laser diode itself. By applying different driving voltages to the comb

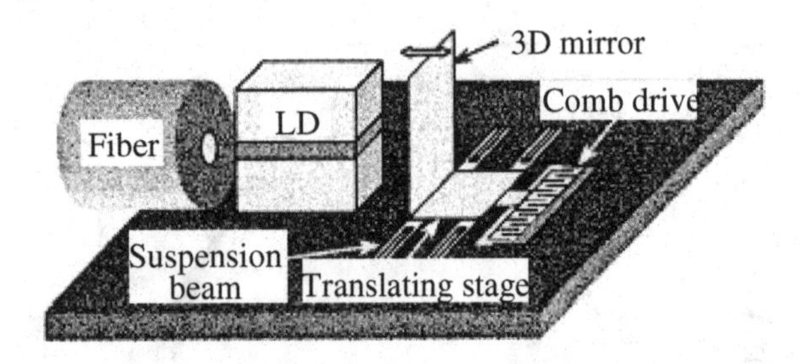

Fig. 5.15 Schematic diagram of tunable laser using a translating vertical micromirror to change the external cavity length. Adapted with permission from *IEEE Photo. Tech. Lett.* **8**, 1, pp. 95-98. Liu et al. (2002).

drive, the 3-D mirror translates and changes the external cavity length, resulting in the tuning of wavelength [Liu *et. al.* (2002)]. As shown in

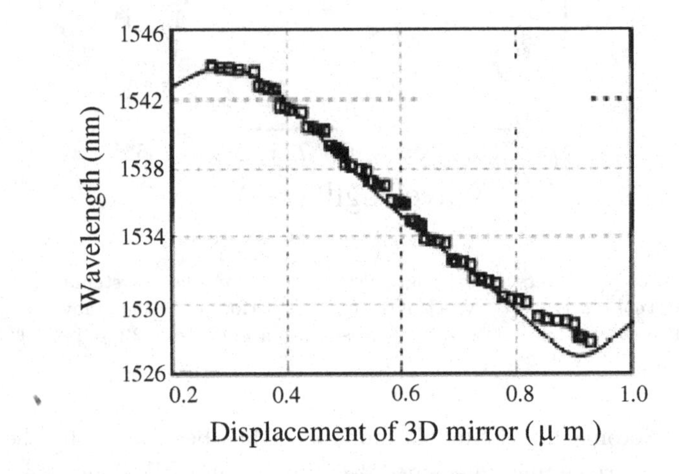

Fig. 5.16 Wavelength variation of the tunable laser with the displacement of 3-D mirror. Solid line is the simulated curve based on experimental data. Adapted with permission from *IEEE Photo. Tech. Lett.* **8**, 1, pp. 95-98. Liu et al. (2002).

Fig. 5.16, a wavelength tuning range of 16 nm is achieved by moving the micromirror laterally by driving an electrostatic comb drive attached to the three-dimensional micromachined mirror. When the 3-D mirror displaces from about 0.3 to 0.9 μm, the wavelength changes from about 1544 nm to 1528 nm.

5.3.2 *ECDL design*

In designing all of the external cavity diode laser system, a few elementary concerns must be taken into consideration. One needs to maximize the feedback, to precisely align the laser cavity to select the wavelength, and to accurately control the laser chip temperature and extract the excess heat [Zorabedian (1996)].

Sufficient feedback strength to the diode laser waveguide mode is required to have stable ECDL operation, to improve the single-mode tunability without mode hopping. The optimal feedback strength depends on the characteristics of the laser and the reflectance of the output facet. The figure of merit for external feedback strength is the cavity loss ratio, i.e., the ratio of the mirror loss of the solitary cavity to the loss of the external cavity. The cavity loss ratio should be at least 20 dB for any external cavity design. To obtain the strong feedback, it is highly desirable to use good quality high numerical objective with numerical aperture of NA$\geq$ 0.55 for collimating lens.

There are two figures of merit for wavelength selectivity. The first one is solitary cavity mode selectivity, which is the ratio of the mode selection filter FWHM bandwidth to the solitary axial mode spacing, it can be written as

$$N_{int} = \frac{\Delta\lambda_{FWHM}}{\Delta\lambda_{int}}. \tag{5.36}$$

Good tunability in the tracking between the oscillation wavelength and the peak feedback wavelength will be obtained with $N_{int} \leq 0.3$ as long as the cavity loss ratio >20 dB. The other one is external cavity mode selection:

$$N_{ext} = \frac{\Delta\lambda_{FWHM}}{\Delta\lambda_{ext}}, \tag{5.37}$$

which is the ratio of the mode selection filter FWHM bandwidth to the external cavity mode spacing. To ensure single mode operation, it is necessary to have $N_{ext} < 1$. Care must be taken to ensure a stable thermal and mechanical structure by using good material and kinematic design principles. Though the ECDLs provide narrow spectral linewidth, they are also much more susceptible to external perturbation than are the solitary lasers that we start with. For stable single-frequency operation, ECDLs need to be isolated from vibrations and pressure fluctuations [Hawthorn *et. al.* (2001)].

5.3.3 *ECDL alignment*

Figure 5.17 depicts the layout of external cavity diode laser of Littrow configuration, which requires a stable thermal and mechanical structure by using good material and kinematical design principle. ECDL consists of a stainless steel baseplate (not shown) that acts as a rigid backbone for laser resonator and heat sink. A Sharp diode laser with antireflection coating is mounted in a small copper fixture that bridges the Peltier cooler and attached to laser baser mount. The base mount is then fixed onto the baseplate. The collimating lens is connected to the laser mount by a stiff spring-steel flexure that is clamped, the lens is mounted in the eccentric ring that is clamped in place after initial coarse alignment. Fine adjustment of focus is done with high-quality fine-pitched screw. A high quality dielectric mirror is glued on the mirror mount with fine-pitched screw. A piezoelectric transducer (PZT) is sandwiched between the mirror and mount. The grating is mounted in the Littman configuration in order to couple the first diffraction order back into the laser diode. A fraction of beam incident on the grating is reflected out of the resonator and constitutes the output of laser. The grating used in this setup has 1800 lines/mm for λ=780 nm. Typical feedback power ratio from 5 to 50 % is coupled into first diffraction order, the grating is glued on a aluminum which is mounted on the baseplate.

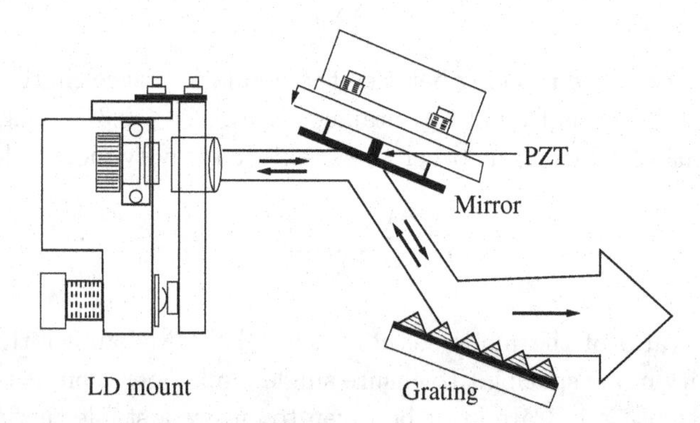

Fig. 5.17 Schematic diagram of a simple mirror-tuned external cavity diose laser system, which consists of three basic elements: laser mount, mirror and grating. The diode laser, Peltier cooler, and collimator are all integrated together in a stable laser mount. LD: laser diode, PZT: piezo tube.

The optical feedback for the purpose of narrowing the line and stabilizing the diode laser has been discussed previously [Fox *et. al.* (1997)]. A diode laser tends to lase at mode frequency with greatest net gain. Fig. 5.18 shows the various contributions to the laser gain profile: i) the laser medium, this depends on the properties of the diode laser material, in particular the bandgap. The gain from this contribution shows a broad spectrum, whose peak depends mainly on laser chip temperature. ii) The internal cavity, the diode junction forms a small etalon, which continues to affect the gain curve even after one of the facets is AR coated. The cavity gain function shifts in frequency with changes in the diode temperature and current. As the temperature is increased, the peaks of both the medium gain and the internal cavity gain curves shift to longer wavelengths. However, they do not shift at the same rate, and the result is that the laser mode hops to different peaks of the cavity gain function. For this reason a typical uncoated diode laser without feedback cannot be tuned to any arbitrary wavelength. iii) The gating feedback, assuming the diffraction limit of the grating, the spectral width $\Delta\nu$ of the grating feedback in the first order will be given approximately by $\nu/\Delta\nu = N$, where ν is frequency and N is the number of grating lines subtended by the laser beam. Usually $\Delta\nu$ could be from dozens to hundreds of GHz. The position of the peak is determined by the grating position. iv) The external cavity, the back facet of laser and the grating consists of the external cavity, whose curve shifts by moving the grating position. One free spectral range is tuned by moving the grating one wavelength. The grating rotates about a pivot position as it translates, which shifts the grating's gain function so its peak partially follows the external cavity peak. This increases the laser's tuning range.

The optical alignment of external cavity laser requires precision and skills. There are several methods to see optical feedback and get the desired single mode frequency ν_0, the gain from each of components should peak at ν_0 as shown in Fig. 5.18. Here are two commonly used ways to achieve the optical feedback.

5.3.3.1 *Threshold current*

We begin with setting the injection current so that the diode laser operates slightly below threshold. By adjusting fine screws, the two fluorescence beams emerging from the grating, (one beam is much weaker than the main one because it has made a complete round-trip inside the laser cavity,) are brought together until they are roughly overlapped. This should have a sudden increase of the brightness of the emission from the lasing. The

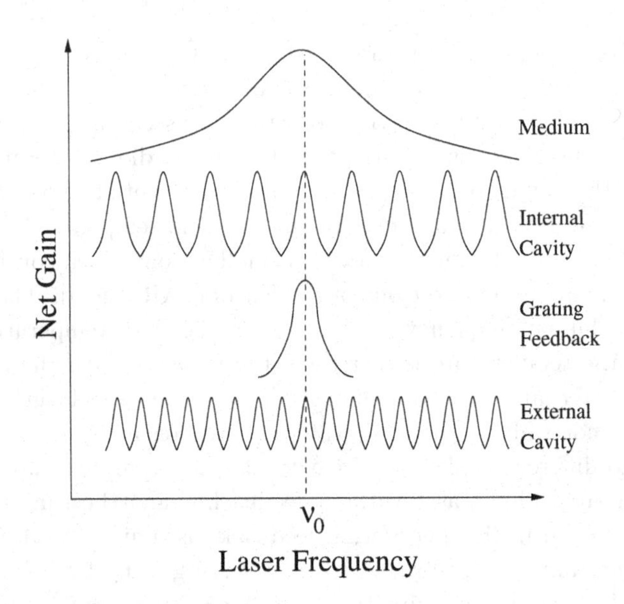

Fig. 5.18 Schematic representation of the various optical elements of external cavity diode laser gain.

injection current is then reduced until the lasing action disappears and alignment is repeated. A sufficiently good alignment is accomplished by decreasing the threshold current by about 10~15 % as compared to the threshold current of free-running diode.

5.3.3.2 *Output power*

We start off with monitoring the output power as a function of the swept injection current, a triangular ramp is applied to the injection current. Monitoring the output power of the ECDL with a large-area photodiode and an oscilloscope obtains L-I curve with optical feedback as shown in Fig. 5.19(a). When there is feedback from the extended cavity, there will be abrupt discontinuous changes in threshold behavior. Iterative adjustment of the focus and extended-cavity alignment will result in the graph as shown in Fig. 5.19(c). The second method is simple and the most often used to align the ECDLs. Also one can use CCD camera to monitor the change of the intensity to get the optimized feedback.

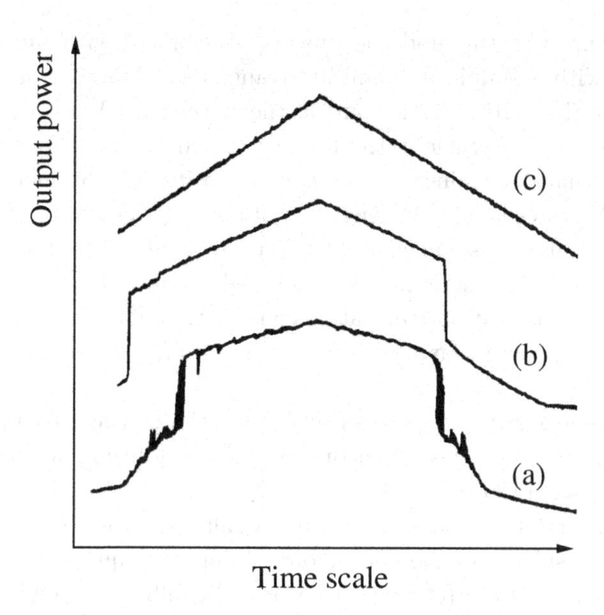

Fig. 5.19 Output power of ECDL versus injection current, which is swept near threshold with a triangle wave. Three L-I curves are shown for different cavity alignments. Curve (a) poor cavity alignment and multimode operation, (b) abrupt jump indicating the single-mode operation, (c) optimum alignment.

5.4 Geometry for mode-hop suppression

In this section we describe the geometries that allow for the single-mode diode lasers scanning for configurations of Littrow and Littman-Metcalf, respectively.

5.4.1 *Littrow configuration*

Mode-hop suppression in a tunable laser by employing a Littrow grating can be obtained in terms of simultaneous sweeping the Littrow grating angle and external-cavity length. The simplest way to obtain such coupled movements is to rotate the Littrow grating about a particular axis. It has shown that the optimal rotation point can been found to obtain a maximal continuous tuning range [de Labachelerie and Passedat (1993)]. It is possible to tune a laser wavelength in a very wide ranges of 240 nm near 1450 nm with an external cavity diode laser [Bagley *et. al.* (1990); Tabuchi (1990)], but with mode-hops. However, a continuous tuning range of 15 nm around 1300 nm without mode hop has been achieved with same type of ECDLs and with a simple mechanical arrangement [Favre *et. al.* (1986)]. In such lasers, the grating angle controls the wavelength λ_r associated with minimum losses. Nevertheless, the lasing wavelength also depends on the cavity length that determines resonant mode positions. Continuous tuning is obtained if the resonant wavelength λ_q of the longitudinal mode number q and the minimum loss wavelength λ_r are spectrally shifted at the same rate to keep the lasing mode in a low-loss region. Such a condition can be fulfilled by use of a rotation-translation combination of the grating position in despite of complex mechanical setup and stability [Favre and Le Guen (1991)].

One can tune a single-mode laser by moving the grating. To understand how the frequency changes when one moves the grating, we investigate several cases [Levin (2002)]. In Fig. 5.20(a), the grating is moved along the laser beam direction, the standing wave oscillating inside the cavity will be stretched, resulting in the continuous tuning of frequency. But as the frequency changes, this stretching also varies the diffraction angle from the grating, after a while, a mode with one more half-wave period inside the cavity will be pointing more directly toward the mirror; i.e., the mode will have low losses, leading to a abrupt mode hop back in frequency as shown in right side of Fig. 5.20(a). If, however, the grating is moved perpendicular to the beam as illustrated in Fig. 5.20(b), the distance between any particular groove on the grating and the mirror does not change, which means that even the cavity is becoming longer, there is no change in frequency. One can also vary the frequency of the laser by changing the angle of the grating, thereby choosing the frequency feedback to the mirror. If the grating is turned around the center point of the beam, as shown in Fig. 5.20(c), the frequency of the cavity at the middle of the beam will not change. This means there is no variation of frequency until the next possible mode has lower losses, and then the laser mode hops to the new frequency.

How should the grating be moved if one wants to achieve the continuous tuning with no mode hops. It is possible to use a grating rotation only to optimize the exact position of the rotation axis to obtain optimum continuous wavelength tuning. The schematic diagram of ECDL is shown in Fig. 5.21, The intersection of the laser axis and the grating plane is denoted by G, the origin of the axis at point O is defined by OG=L, L is optical length of the laser cavity. The first diffraction order of the grating is reflected back to the cavity for a grating angle θ. For a grating period d,

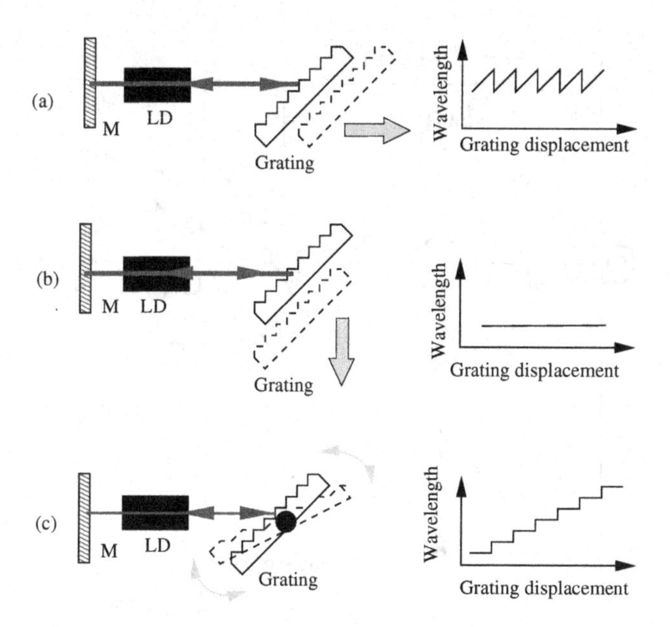

Fig. 5.20 Different ways to move the grating and its corresponding change in wavelength. Adapted with permission from *Opt. Lett.* **27**, 4, pp. 237-239. Levin (2002).

λ_r is given by

$$\lambda_r = 2d \sin\theta. \tag{5.38}$$

The resonant cavity-mode wavelength λ_q is given by well known expression λ_q=2L/q, taking into consideration grating translation in its plane. One finds λ_q,

$$\lambda_q = \frac{2L}{q + t_0(\theta)/d}, \tag{5.39}$$

where $t_0(\theta)$ is the distance from G to a grating groove point that is taken as reference. It is assumed that the grating is rotated about an axis $R(z_0, y_0)$ parallel to the grating lines and that before any grating rotation its initial angle is θ_0, the cavity optical length is $L(\theta_0) = L_0$, and the q-th mode frequency is exactly at the minimum-loss frequency λ_r:

$$\lambda_q(L_0) = \lambda_r(\theta_0). \tag{5.40}$$

When the grating is rotated the angle θ varies and thus λ_r is shifted. One

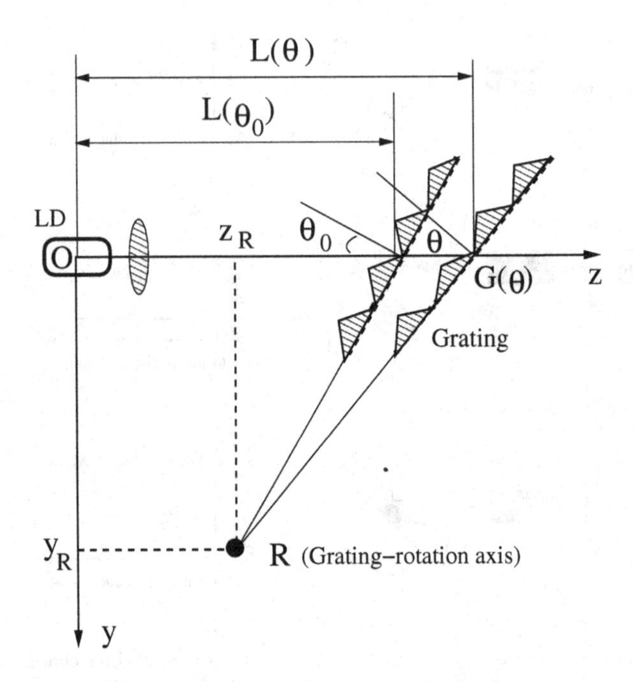

Fig. 5.21 Schematic diagram of laser configuration. Adapted with permission from *Appl. Opt.* **32**, 3, pp. 270. Labachelerie and Passedat (1993).

is interested in the difference,

$$F(\theta) = \lambda_q(\theta) - \lambda_r(\theta), \tag{5.41}$$

which is zero for $\theta = \theta_0$ and must remain zero as long as possible for a large grating rotation. If $F(\theta)$ keeps smaller than half the cavity-mode spacing $\Delta\lambda$, the q−th mode will stay dominant and no mode hop will occur while the wavelength is tuned. Therefore, the condition for mode-hop suppression is

$$|F(\theta)| < \Delta\lambda/2, \tag{5.42}$$

this condition is obtained for $\theta = \theta_0$ and can be maintained for large $(\theta - \theta_0)$ values if the derivatives of $F(\theta)$ vanish at $\theta = \theta_0$.

By use of Eqs. (5.38) and (5.39), Eq. (5.41) can be written as

$$F(\theta) = \frac{2L(\theta)}{q + t_0(\theta)/d} - 2d\sin\theta. \tag{5.43}$$

For simplicity, it is assumed that for $(\theta = \theta_0)$ the reference-grating facet is exactly on the laser axis $[t_0(\theta_0) = 0]$ and that there is one resonant mode at the minimum-loss wavelength $F(\theta_0) = \lambda_q(\theta_0) - \lambda_r(\theta_0)$. Then $F(\theta)$ can be written as:

$$F(\theta) = 2d[\frac{L(\theta)\sin\theta_0}{L_0 + t_0(\theta)\sin\theta_0} - \sin\theta], \tag{5.44}$$

to determine the optimal rotation points, one takes the first derivative of $F(\theta)$,

$$F'(\theta_0) = 2d[\frac{L'(\theta_0)\sin\theta_0 - t_0'(\theta_0)\sin^2\theta_0}{L_0} - \cos\theta_0]. \tag{5.45}$$

To the first order near $\theta = \theta_0$ we have

$$t_0(\theta_0) = [\frac{L(\theta_0) - z_R}{\cos\theta_0}](\theta - \theta_0), \tag{5.46}$$

which gives

$$t_0'(\theta_0) = [\frac{L_0 - z_R}{\cos\theta_0}], \tag{5.47}$$

$L'(\theta_0)$ can be calculated from the expression for $L(\theta)$, which is obtained from geometry of Fig. 5.21, that is given by

$$L(\theta) = -y_R[\frac{\sin\theta_o}{\cos\theta} - \tan\theta] - z_R[\frac{\cos\theta_o}{\cos\theta} - 1] + L_0\frac{\cos\theta_o}{\cos\theta}, \tag{5.48}$$

it is then pretty easy to find that $F'(\theta_0) = 0$ for

$$y_R = \frac{L_0}{\tan\theta_0}. \tag{5.49}$$

Therefore, the optimum rotation point that provides continuous tuning at the first order is the point of the line R_1R_2 as indicated in Fig. 5.22. If one now makes an expression of $F(\theta)$ to the second order for the point of the line R_1R_2 and uses the value of y_R by Eq. (5.49), one can obtain

$$F(\theta) = d\sin(\theta_0)(\frac{z_R}{L_0} + 1)(\theta - \theta_0)^2, \tag{5.50}$$

which cancels out for $z_R = -L_0$. Thus the optimum rotation point that provides continuous tuning to the second order of $F(\theta)$ is R_4, as shown in Fig. 5.22.

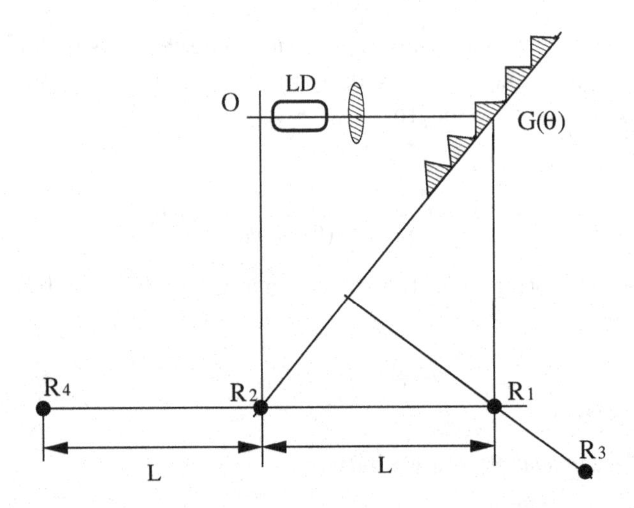

Fig. 5.22 Particular rotation axes: R_4 is optimum rotation point. LD: laser diode, G: grating. Adapted with permission from *Appl. Opt.* **32**, 3, pp. 270. Labachelerie and Passedat (1993).

5.4.2 *Littman-Metcalf*

We here describe a geometry that allows for the single-mode scanning of diode lasers for Littman-Metcalf configuration, It has been discussed that the right choice of a grating rotation axis can provide changes simultaneous in cavity length and diffraction angle that exactly match the requirements needed for continuous single-mode scanning.

The basic geometry of the single-longitudinal-mode grazing-incidence diode laser is shown schematically in Fig. 5.23 as proposed by Liu [Liu and Littman (1981)] and [Shoshan *et. al.* (1977)]. Tuning is accomplished simply by rotating a mirror. By carefully selecting the position of pivot point about which the tuning mirror is rotated. One could scan simultaneously the cavity length and grating feedback angle, therefore permitting a continuous single-mode scan over a limited range. It has found possible to define a pivot point that satisfies this tracking condition exactly over the entire tuning range of the grating.

In order to achieve tracking it is necessary to satisfy the following two equations as in the Littrow configuration, which together determine the laser wavelength:

$$\lambda_r = \frac{d}{p}(\sin\theta_0 + \sin\phi), \tag{5.51}$$

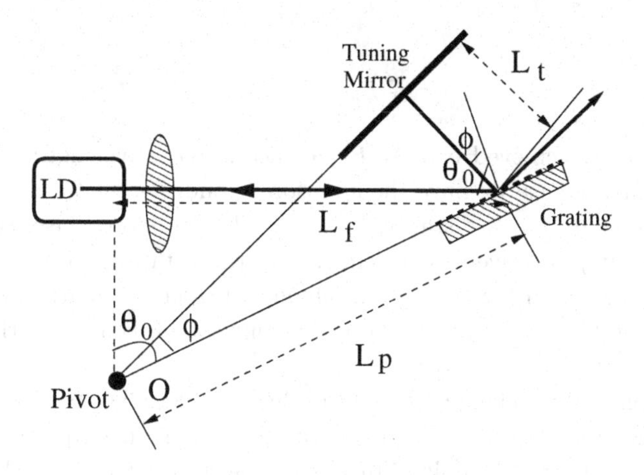

Fig. 5.23 Grazing-incidence pulsed dye laser with self-tracking geometry. Adapted with permission form *Opt. Lett.* **6**, 3, pp. 117-118. Liu and Littman (1981).

and

$$\lambda_q = \frac{2L(\phi)}{q} = \frac{2}{q}(L_f + L_t(\phi)), \tag{5.52}$$

where $L(\phi)$ is the optical cavity length (L_f is the optical distance between the grating) and the fixed mirror. $L_t(\phi)$ is the optical distance between the grating and tuning mirror, q is the mode number, d is grating period, p is the diffraction order, θ_0 is the incidence angle, and ϕ is the diffraction angle. If one chooses the pivot axis to be the intersection of the surface planes of the grating and tuning mirror, and designates the distance between the grating angle center and pivot axis by L_p, then $L(\phi)$ is just

$$L(\phi) = L_f + L_p \sin\phi, \tag{5.53}$$

substituting this result into Eq. (5.52) gives

$$\lambda_r = \frac{2}{q}(L_f + L_p \sin\phi). \tag{5.54}$$

Comparison of Eqs. (5.51) and (5.54) reveals that, if

$$\frac{2}{q}L_f = \frac{d}{p}\sin\theta_0, \tag{5.55}$$

and

$$\frac{2}{q} L_p = \frac{d}{p}, \tag{5.56}$$

then tracking is achieved exactly for all accessible wavelength. For the grazing-incidence case we have the additional condition $\sin\theta_0 \approx 1$, so that the pivot is located at a distance $L_p \approx L_f$. The proposed laser-scanning scheme does impose severe requirements on the rotation mount used for the tuning mirror. In particular, a mechanical runout of as little as a half-wavelength will result in a mode hop. Care must be taken to set the pivot location correctly.

Unfortunately, mechanical tuning is slow in all ECDL configurations, and it is difficult to achieve high repetition rates, high frequency tuning speed, and good reproducibility. In order to overcome these problems, we can, instead of moving grating mechanically, tune the frequency of laser electronically. We will explore them more details in the next chapter.

Chapter 6

Implementation of Tunable External Cavity Diode Lasers

In the last chapter, we were concerned with the systems of external cavity diode lasers, and we introduced various configurations of external cavity diode lasers and approaches for achieving single-mode continuous tuning in external cavity diode lasers. In this chapter, we demonstrate the various ways to implement external cavity diode laser systems. Widely continuous tunable external cavity diode lasers are produced by mechanically and micro-electro-mechanically tuning the external mirrors or gratings simultaneously. Electronically tunable external cavity diode lasers are produced by use of acousto- and electro-optic tunable filters and liquid crystal spatial light modulators. Finally, we develop some special tunable external cavity diode lasers such as blue-violet and high powered ones.

6.1 Widely continuous tunable ECDLs

A continuous wavelength tunability of 82 nm is described in this section by mechanically tuning the mirror and grating simultaneously. Wide tunable external cavity diode lasers based on micro-electro-mechanical-system in different forms are introduced as well.

6.1.1 *Mechanical wide tuning*

As described in the previous chapters, if the wavelength tuning is achieved by simply rotating the grating without translating it, external-cavity mode hopping should occur over about the lasing cavity free spectral range, which is typically 3 GHz for a 5- cm-long cavity. Continuous tunability without longitudinal mode hopping can be obtained with ECDLs by combined rotation-translation of the grating, as shown in Fig. 6.1, synchronous rotation and translation of the tuning mirror eliminates mode-hopping. The first demonstration of this technique has been reported with a range of 15 nm without mode hopping around 1226 nm wavelength [Favre *et. al.* (1986)].

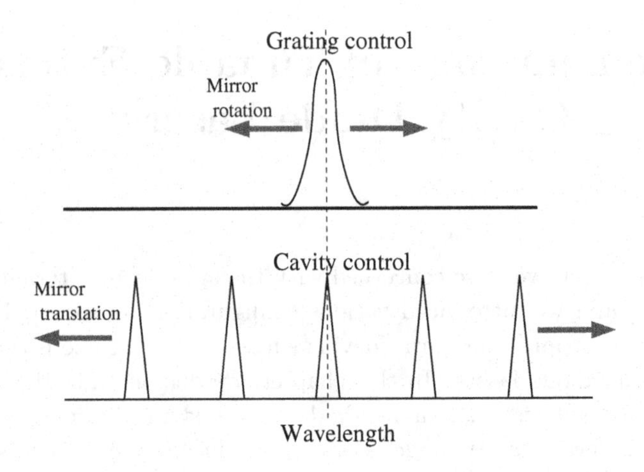

Fig. 6.1 Schematic of continuous tuning with combination of rotation and translation. Adapted with permission from *Laser Focus World*, June. Lang (1998).

An improved ECDLs operating around 1540 nm with a wavelength tunability of 82 nm with mode hopping has been demonstrated [Favre and Le Guen (1991)], the configuration is schematically shown in Fig. 6.2(a). The 1.55 μm buried heterostructure semiconductor laser is AR-coated on one facet with a residual reflectivity decreasing from 10^{-4} at λ=1550 nm to 10^{-5} at λ=1582 nm. The output of the AR-coated facet is collimated on a 1200 groove/mm diffraction grating by a 0.615 NA lens with a 6.5 mm focal length, the optical length of the ECDLs cavity is 50 mm at λ=1540 nm, The output is isolated by an optical isolator with 60 dB extinction ratio. The grating rotation around the central ruling A is mechanically controlled by a push rod, the end of which is freely moving along the CB axis. The rotating grating is mounted on a motorized translator which controls the simultaneous change of the cavity length L and of the grating incidence angle θ.

Single-mode emission is obtained when the wavelength λ_q of an ECDL mode matches the wavelength λ_r of the grating in retroreflection conditions. The differential micrometer screw shown in Fig. 6.2(a) is used for fine adjustment of λ_r by slightly rotating the grating without changing the cavity length L. The wavelength detuning over which the ECDL emission is locked on a longitudinal mode has been experimentally found to be equal to the free spectral range of the ECDL cavity. The continuous tuning condition is satisfied if λ_r and λ_q can be shifted at the same rate which requires that $AB = L/\sin\theta$ is constant over the tuning range as shown in Fig. 6.2(b). The mechanical tuning procedure consists of adjusting the CB axis posi-

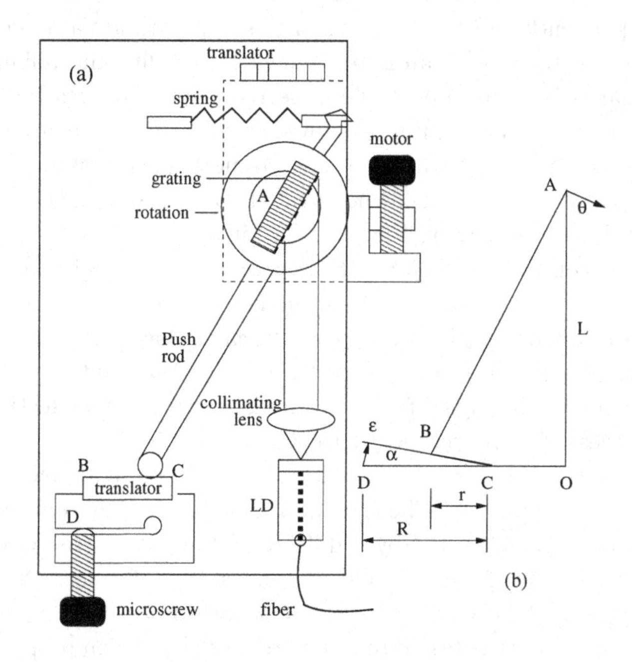

Fig. 6.2 Configuration of external cavity lasers. (a) Arrangement, (b) equivalent geometry scheme, L=50 cm, r=7 mm, R=24 mm. Adapted with permission from *Electron. Lett.* **27**, 2, pp. 183. Favre and Le Guen (1991).

tion with the micrometer screw to maximize the continuous tuning range without mode hopping. The maximal continuous tuning range $\Delta\lambda$ can be calculated as a function of the micrometer detuning $\delta\epsilon$, which is given by

$$\Delta\lambda = \frac{R\lambda^2}{2\delta\epsilon(L\tan\theta - r)}, \tag{6.1}$$

where λ is the mean wavelength and r and R are mechanical distances defined in Fig. 6.2(b).

6.1.2 *Micro-electro-mechanical-system wide tuning*

External cavity diode lasers based on conventional stripe-waveguide semiconductor diode lasers and piezoelectrically actuated blazed grating were originally developed in the late 1980s by BTRL and CNET. More recently, semiconductor lasers have been combined with mirrors and gratings on micro-machined electrostatic actuators by NEC, Iolon, Fujitsu, Santur, and Agility, independently. The advantages of this approach are a reduction in size and cost, combined with an improvement in stability and reliability.

Traditional ECDLs have been piezo-electrically tuned with bulk opto-mechanical assemble. A new and high power, widely tunable micro-external cavity diode laser based on a micro-electro-mechanical-system (MEMS) electrostatic actuator has been demonstrated [Berger and Anthon (2003)] A schematic diagram shown in Fig. 6.3 is a Littman-Metcalf external cavity diode laser tuned by a silicon MEMS actuator. The light source is a InGaAsP/InP multi-quantum-well laser diode with high output power and wavelength at 1.55 μm. Reflection of the intracavity facet of the laser diode is suppressed by use of low reflectance ($< 2 \times 10^{-3}$) antireflection coating in combination with an angled facet, this makes possible and effective facet reflectance of less than reflectance $< 10^{-4}$.

The laser output beam is collimated at the opposite diode facet, which serves as the output coupler of the resonator. Light emerging from the intracavity diode facet is collimated by a diffraction-limited, micro-optical lens and then diffracted at grazing-incidence from a 50% efficiency, free-space diffraction grating. The directly reflected, zeroth order beam constitutes the laser output, while the first diffracted order of laser beam propagates to an external mirror mounted out-of-plane on a rotary silicon micro-actuator. Wavelength tuning is achieved by applying a voltage to the micro-actuator, which rotates the mirror to allow a specific diffracted wavelength to couple back into the laser diode. The actual wavelength of the laser output is determined by the gain bandwidth of the diode, the grating dispersion and the external-cavity mode structures. The laser is tuned by applying voltage to the comb elements of the MEMS actuator to produce an electrostatic force that rotates the mirror about its virtual pivot. The angular range of the actuator determines the tuning range. A variety of 150 V actuators, with ranges of up to $\pm 2.8\,^{\circ}$, have been used to tune over a wavelength range of up to 42 nm. The mirror's geometrical pivot can be designed to adjust the cavity length. The laser wavelength determined by the diffraction angle then scans synchronously with the grating filter passband, and the laser

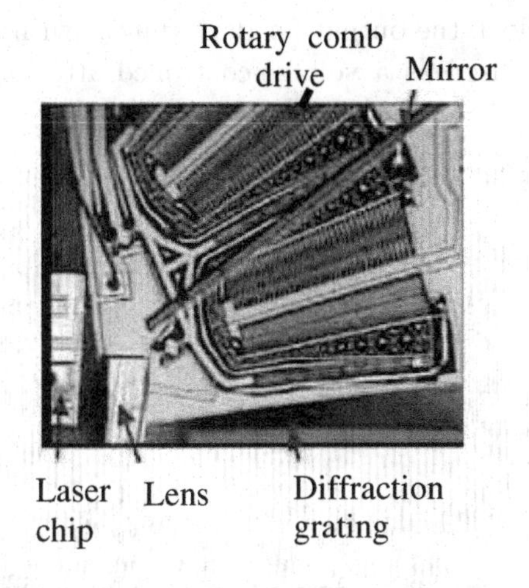

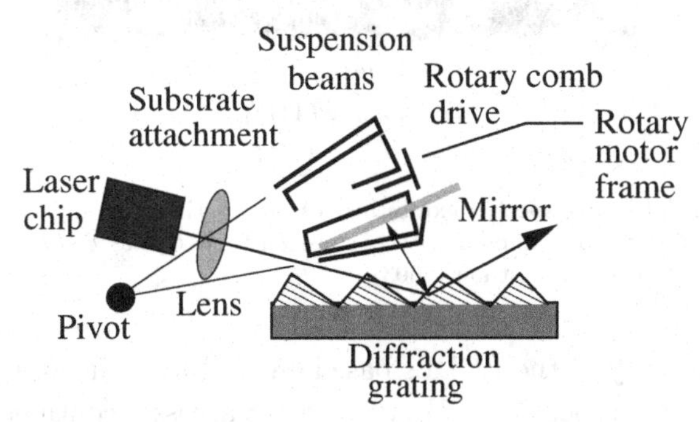

Fig. 6.3 Micro-ECDL photograph and schematic Littman-Metcalf cavity configuration with a laser diode, collimating lens, diffraction grating, and external mirror mounted on a MEMS rotary actuator. Mode-hop-free tuning is achieved by rotating the mirror about a virtual pivot point defined by the intersection of two lines extending from the actuator suspension beams. Adapted with permission from *Optics & Photonics News*, March, pp. 43-49. Berger and Anthon (2003).

tunes continuously without mode hops. The actuator voltage determines the laser frequency with an open-loop accuracy of approximately 10 GHz, and the frequency is then stabilized to ± 1.25 GHz by use of the error signal from wavelength locker etalon in a servo that adjusts the mirror position.

The collimated output beam from the ECDLs, with output power of up to 70 mW, passes through an isolator and is coupled into a polarization-maintaining (PM) fiber pigtail with 65% coupling efficiency. A beam splitter reflects 5% of the light into an integrated wavelength locker. The MEMS shutter is used to block the output for dark tuning, with suitable external feedback, it can be used as a voltage-controlled attenuator (VOA). The

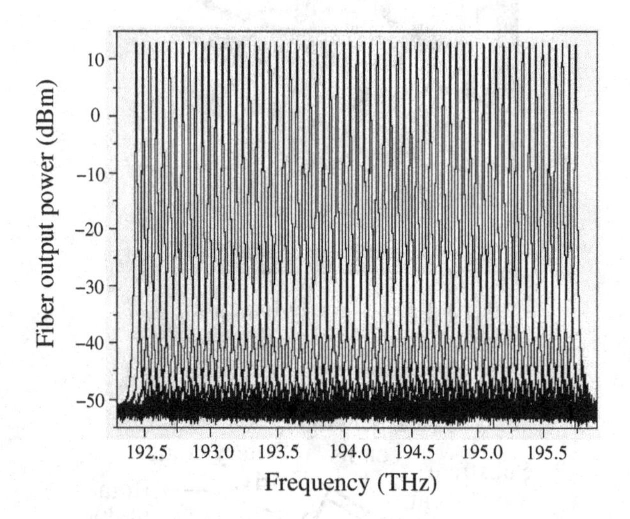

Fig. 6.4 Micro-ECDl lasing spectra exhibiting 40 nm continuous range with >50 dB side mode suppression ratio. Adapted with permission from *Optics & Photonics News*, March, pp. 43-49. Berger and Anthon (2003).

wavelength tunability of the ECDLs based on MEMS is shown in Fig. 6.4, which presents a superposition of lasing spectra across a continuous 40 nm tuning range. Peak output power of up to 40 mW has been achieved from this ECDLs. An external cavity laser can be also fabricated entirely on one chip, using an etched grating in a semiconductor.

An external cavity diode lasers based on stripe waveguide semiconductor optical amplifier, a ball lens, and a blazed diffraction grating fabricated by vertical deep reactive ion etching of silicon has been demonstrated recently [Lohmann and Syms (2003)]. The fabrication process of grating is very simple, a 3 μm thick layer of resist is deposited, patterned into a staircase layout by contact lithography, and hard-baked to act as a surface mask. The silicon is then etched using a cyclic etch-passivation process. After etching, residual resist is stripped in a plasma asher, reflectivity is then enhanced by use of a sputtered layer of Au. Bandpass characteristics for the 12-th order grating has been given, tuning of an external cavity laser over a 120 nm spectra range has been demonstrated, with a maximum single-mode fiber-coupled power of 1 mW and side-mode suppression ratio of 30 dB.

A novel configuration for a wavelength tunable laser [Pezeshki *et. al.* (2002)] has been proposed and demonstrated, that provides wide tuning, distributed feedback (DFB) performance, and reliability at a DFB laser array, a micro-mechanical mirror has been used to select one element of the array. The MEMS tilt loosens the tolerances since the fine optical alignment is done electronically. Only one laser is operated at a time, with coarse tuning realized by selecting the correct laser and fine tuning by adjusting the chip temperature.

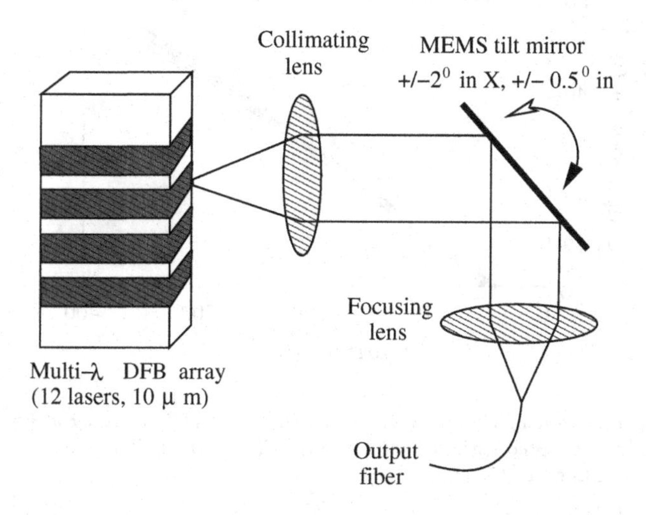

Fig. 6.5 Schematic of tunable laser package. The MEMS tilt mirror in the focal plane of the collimating lens selects one laser from the DFB array and allows for electronic fine tuning of alignment. Adapted with permission from *IEEE Photon. Tech. Lett.* **14**, 10, pp. 1457-1459. Pezeshki et al. (2002).

The schematic diagram is shown in Fig. 6.5. Rather than a complex and lossy integrated combiner, an external MEMS tilt mirror is used to select the appropriate DFB, the mirror is placed at the focal plane of the collimator lens, and thus corrects for the spatial variation of the generated beam. The approach has the advantage of the tremendously simplifying the chip, since no active/passive translations are needed, and chip can be made about the same size as the standard fixed wavelength DFB. The optical loss of passive combiner is eliminated with nearly full output power of the laser coupled out to the fiber.

The package is simple to manufacture and mechanically robust since the time-consuming and costly fine alignment is now done electronically with the tilt mirror. A feedback loop maintains the optimum alignment and compensates for any possible mechanical drift and creep. The DFB array chip contains 12 lasers at a 10 μm pitch with a wavelength spacing of 2.8 nm. The lasers share the same strained quantum well InGaAsP gain medium and only vary in grating pitch, it can deliver about 50 mW chip power at 300 mA current at case temperature of 30 °C, the light-current curve of such an array is illustrated in Fig. 6.6. With such a closely spaced

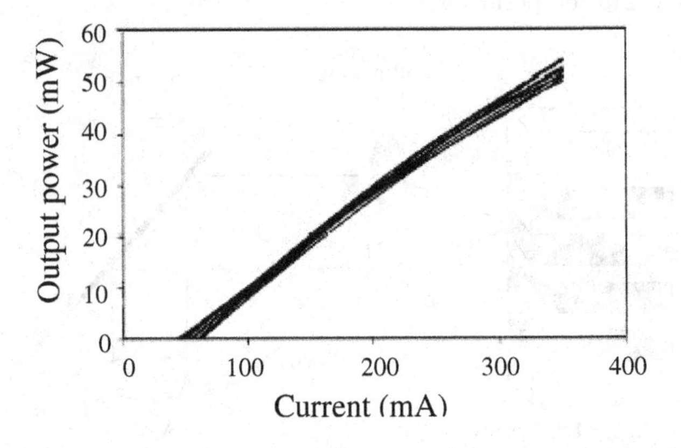

Fig. 6.6 Superimposed light current curves of a 12-element laser array at 35 °C heat sink temperature. Adapted with permission from *IEEE Photon. Tech. Lett.* **14**, 10, pp. 1457-1459. Pezeshki et al. (2002).

array, only a small deflection of the MEMS mirror is needed, which in this case corresponds to about ±1.5 °. But the mirror in fact is designed to tilt ±2 ° in the plane and also tilt ±0.5 ° in the other axis, thus allowing coarse position of the positions. The butterfly package is combined with control electronics and an external wavelength locker, The 33 nm total tuning at 20 mW fiber coupled power is obtained in a fully functional module. Fig. 6.7 shows superimposed spectra at 50 GHz ITU channels, tuning time is typically a second between channels, the module is locked on to the ITU grid within 0.4 GHz.

6.2 Electronically tuning external cavity diode laser

Various configurations using electronically controlled acousto- and electro-optic tunable filters [Harris and Wallace (1969)], as well as spatial light

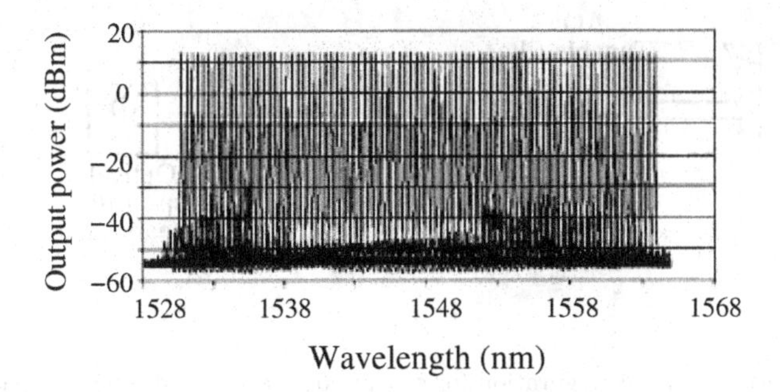

Fig. 6.7 Superimposed spectra of module of 84×50 GHz ITU channel at 12 dBm or 20 mW. Adapted with permission from *IEEE Photon. Tech. Lett.* **14**, 10, pp. 1457-1459. Pezeshki et al. (2002).

modulator have been demonstrated and analyzed in this section. With no mechanical readjustment, these devices are tuned by changing the frequency of the radio frequency (RF) driver and/or scanning voltage applied to the tunable filters which provide tunable wavelength by forward phase-matched coupling between different index modes in a host medium and incident laser beams.

6.2.1 *Wavelength tunability by acousto-optic modulator*

An acousto-optic filter has a narrow optical transmission bandwidth with the center frequency electronically controlled by the drive frequency, an acousto-optic modulator to tune a dye laser in the visible spectral range has been used earlier 1970s [Taylor *et. al.* (1971)]. Fast tuning of diode laser at 850 nm using an acousto-optic modulator inside the laser cavity has been discussed [Coquin and Cheung (1988)]. Electronic tuning of a 1300 nm InGaAsP semiconductor laser over a wide tuning range of 83 nm using an acousto-optic filter in an external cavity configuration was reported [Coquin *et. al.* (1989); Boyd and Heisman (1989)], both single and multi-wavelength operation were observed.

The basic cavity configuration for acoustically tuned, external cavity is shown schematically in Fig. 6.8. The acousto-optic filter devices is made of TeO_2 crystalline material with its FWHM bandwidth 3 nm, and the peak transmission at center wavelength λ_0 is about 80% for acoustic drive power 3 W. Tuning is achieved by varying the frequency of the drive voltage applied to the acoustic transducer, one can tune λ_0 over a range of 1.2 μm

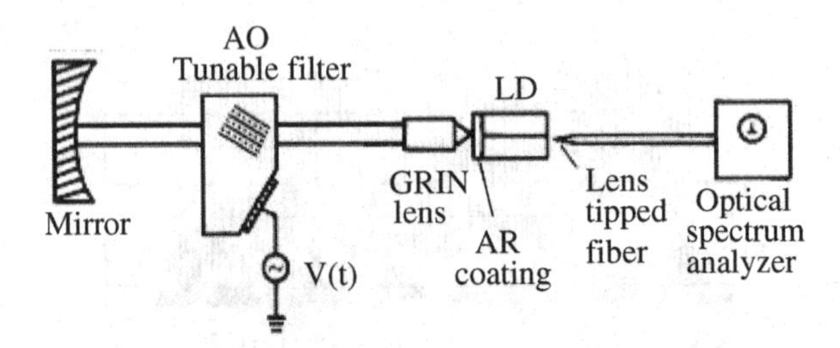

Fig. 6.8 External cavity configuration for acousto-optically tuned semiconductor laser. LD: laser diode, AR: antireflection. Adapted with permission from *IEEE J. Quantum Electron.* **25**, 6, pp. 1575-1579. Coquin et al. (1989).

to 1.6 μm by varying the electric drive frequency from 55 MHz to 75 MHz with the time required to change one wavelength to another being 3 μs for the device. Optical spectra of the tuned optical output of the tunable laser are shown in Fig. 6.9. The upper and lower spectra show the extremes of the tuning range, while the middle trace is a typical spectrum from center of the range, which spans 83 nm. Good mode purity with ~30 dB main mode to side mode suppression ratio is realized over almost this entire range. However, each main mode contains many external cavity modes which are not individually resolved by optical spectrum analyzer.

A unique feature of the acousto-optic filter is that multiple grating can be induced in the crystal by applying simultaneously multiple electric drive frequencies to the transducer input. By carefully adjusting the levels of the signals at the different wavelengths, one can achieve lasing at several different wavelengths simultaneously. The spectra of output light are shown in Fig. 6.10. One unfortunate characteristic of the acousto-optic filter is the fact that an optical wave of frequency ν is shifted up and down in frequency when passing through the filter by an amount equal to the acoustic drive frequency f. This leads to the normal modes of the acousto-optically tuned external cavity laser being time-dependent chirping mode, that is

$$\nu_q = \frac{c}{2l}(q + f_{chirp}t), \tag{6.2}$$

where $f_{chirp} = 2f$ is chirp frequency, a factor of 2 arises from the fact that the laser go and back the filter twice. For many applications, it is nuance that the normal mode of the laser is chirping. The frequency chirping due to the shift can be compensated for by introducing a second acousto-optic modulator inside the cavity, as shown in Fig. 6.11, the second one can easily be configured to provide an equal and opposite frequency shift to that of

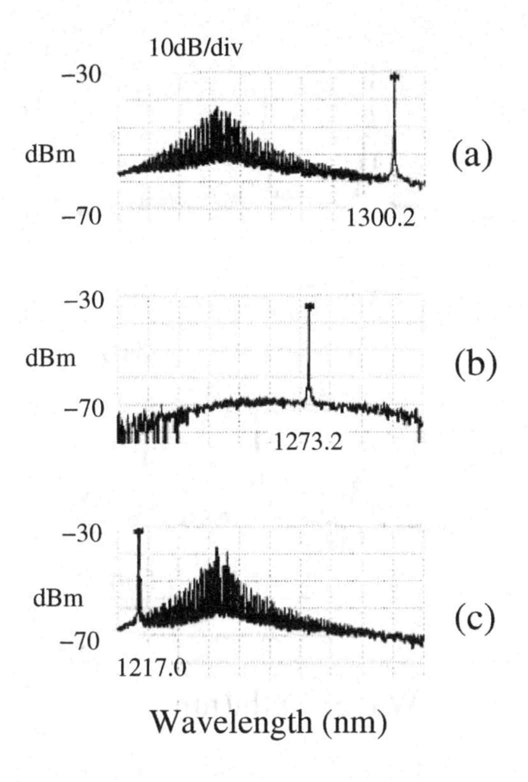

Fig. 6.9 Optical spectrum analyzers of output of acousto-optically tuned laser diode at three values of acoustic drive frequency. viz., (a) f=69.494 MHz, (b) f=71.1 MHz, and (c) 74.57 MHz. Total tuning range shown in the figure is 83 nm. Adapted with permission from *IEEE J. Quantum Electron.* **25**, 6, pp. 1575-1579. Coquin et al. (1989).

the first one, and thus the chirping is eliminated.

Figure 6.12 shows the optical spectra of the laser output obtained with a frequency compensated external cavity using two acousto-optic filters inside the cavity, with both filters driven at same frequency. In this way, one eliminates the chirping. However, the laser is still running in multimode and discontinuously. It has been proposed and demonstrated that a novel approach can achieve rapid and phase-coherent continuous broadband tuning of a single-mode external cavity semiconductor lasers [Kourogi *et. al.* (2000)]. To this end, two acousto-optic modulators have been used to control the angle of incident light on a diffraction grating and the effective round trip optical phase electronically. To have the round trip phase, a long delay line to generate a modulation signal of the acousto-optic device

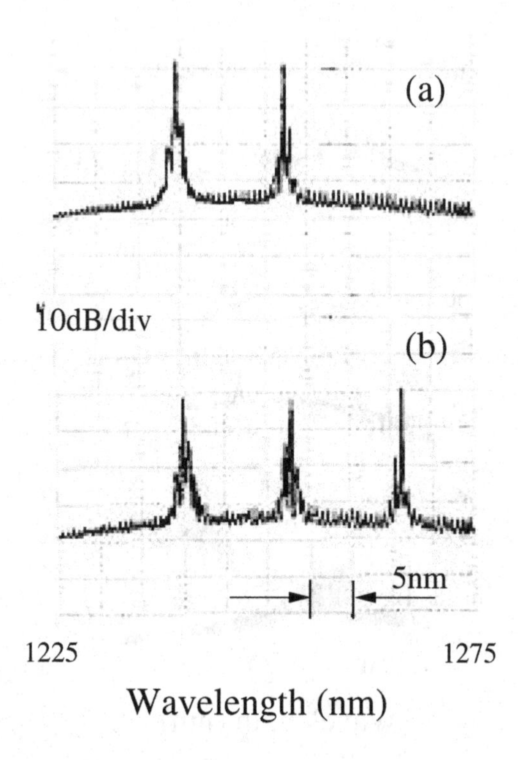

Fig. 6.10 Optical spectrum analyzer traces of output of tunable laser showing multiple wavelength tuning when multiple frequencies are injected into the acousto-optic filter. (a) f=81.3 MHz and 82.3 MHz, (b) f=80.3 MHz, 81.3 MHz and 82.3 MHz. Adapted with permission from *IEEE J. Quantum Electron.* **25**, 6, pp. 1575-1579. Coquin et al. (1989).

is employed.

Figure 6.13 shows the diagram of the proposed electronically tunable external cavity semiconductor laser. The configuration is different from that of a previous one in two ways [Coquin and Cheung (1988)]. First, instead of acousto-optic tunable filter, an external grating and a pair of acousto-optic modulators are used as the wavelength selective element in a Littrow configuration. This allows fast and accurate selection of the lasing wavelength by varying only the incident angle to the grating that interlocks with the drive frequency (f) of the acousto-optic modulator. Each acousto-optic modulator is set in the same diffraction direction. Compared with the previous methods [Coquin *et. al.* (1989)], this approach gives high-resolution wavelength selectivity, enables the single-mode operation of

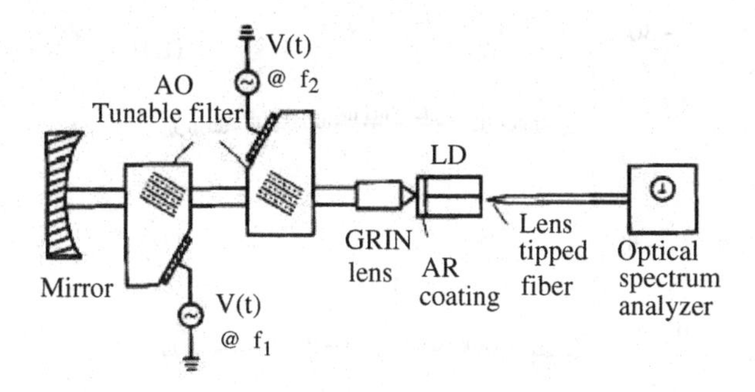

Fig. 6.11 External-cavity configuration using two acousto-optic filters to compensate for optical frequency shift. Adapted with permission from *IEEE J. Quantum Electron.* **25**, 6, pp. 1575-1579. Coquin et al. (1989).

this laser. Second, an electrical delay line is incorporated into the line for microwave signal going to one of the acousto-optic devices. The basic idea is similar to that in an external cavity laser, in which continuous tuning over a broad wavelength range has been realized by simultaneous mechanical control of the external-cavity length and the oscillation wavelength of the laser [de Labachelerie and Passedat (1993)].

The continuous tuning of laser frequency is achieved by simultaneously changing both the electric drive signal frequency f and the relative phase ϕ. Fig. 6.14 illustrates the continuous laser frequency tuning. When the electric frequency f is continuously varying without changing its phase, the laser experiences a stairway-shaped mode hop through every Fabry-Perot mode of the external cavity, as shown in Fig. 6.14(a), since one cannot change the frequencies of the external cavity modes by changing f. When one continuously changes ϕ without varying the f, the laser experiences a saw-shaped mode hop for every π, as displayed in Fig. 6.14(b). This is a so-called chirped-frequency operation in a frequency-shifted feedback laser that was studied by Nakamura [Nakamura *et. al.* (1997)]. This operation explains how one can tune the laser frequency while avoiding mode hop when f and ϕ are changing simultaneously as depicted in Fig. 6.14(c).

The simplest method of varying both f and ϕ simultaneously is to use an electrical delay line, as shown in the inset of Fig. 6.13, one can vary the relative phase ϕ of the electrical drive signals applied to the acousto-optic devices by changing f in terms of the relation of $\phi = \pi f \tau$, where τ is the delayed time. Hence the effective round trip phase shift of the optical signal that returns to the laser is 2ϕ. Therefore the frequency of the Fabry-

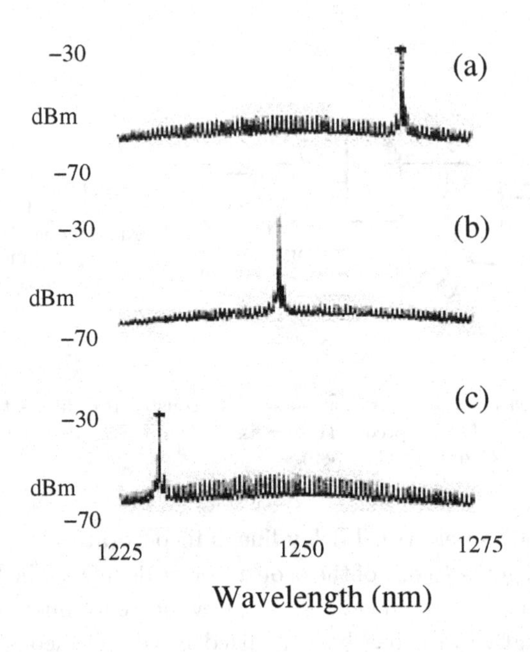

Fig. 6.12 Optical spectrum output of non-chirping tunable laser using two acousto-optic filters inside the cavity. (a) f=68 MHz, (b) f=69 MHz, and (c) 70 MHz. Adapted with permission from *IEEE J. Quantum Electron.* **25**, 6, pp. 1575-1579. Coquin et al. (1989).

Perot modes of an external cavity changes its free spectral range (FSR) by changing ϕ by π. To match the frequency tuning of the laser by f and ϕ, one can obtain the optimum delay time from $\tau = \langle df_{laser}/df \rangle/(2FSR)$, where $\langle df_{laser}/df \rangle$ is the tuning of the laser frequency to the modulation frequency. In Fig. 6.14(a), $\langle df_{laser}/df \rangle$ can be referred to as the coefficient of the center frequency of the passband of the grating feedback to the modulation frequency. Thus it is possible to tune the laser frequency continuously by use of an electrical delay line as a phase shifter to vary both f and ϕ simultaneously.

An InGaAs Fabry-Perot semiconductor cavity with antireflection coating ($< .1\%$) at 1.55 μm on the facet was used to demonstrate the continuous tuning of laser frequency via the technique described above. The diffraction efficiency of the acousto-optic modulators was $\sim$ 70%. A grating of 1200 grooves/mm was used. Since the diffraction efficiency of the grating was 70%, the total reflectivity from the acousto-optic modulators and the grating was 16%. Single-mode operation and electrical tuning over 2 nm

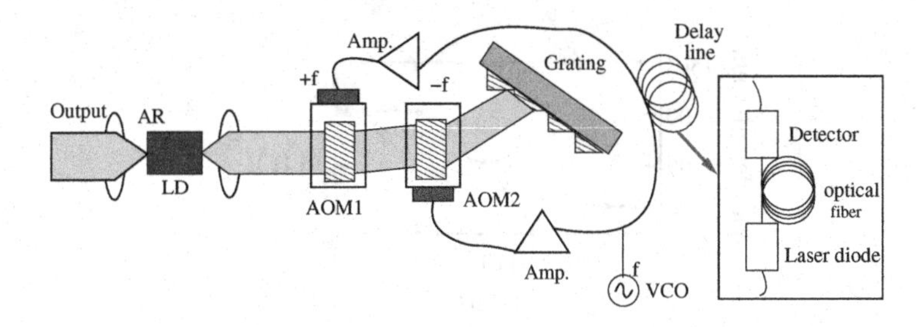

Fig. 6.13 Electronically tunable external-cavity semiconductor laser structure: AOM1, AOM2: acousto-optic modulator; VCO, voltage-controlled oscillator; Amp. microwave amplifier; LD: laser diode. Adapted with permission from *Opt. Lett.* **25**, 16, pp. 1165. Kourogi et al. (2000).

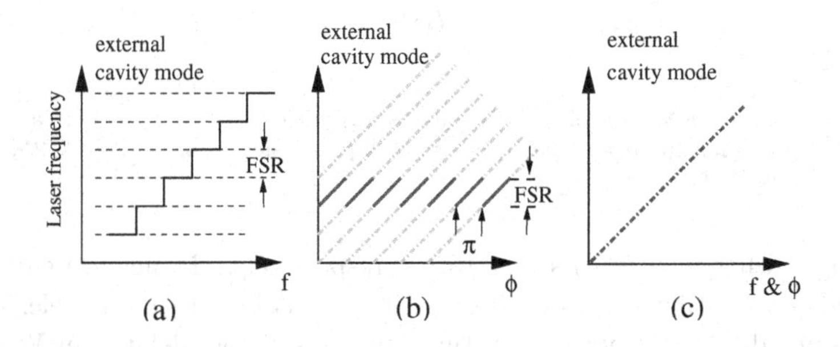

Fig. 6.14 Explanation of the laser frequency tuning: (a) f is changed without changing ϕ. (b) ϕ is changed without changing f. (c) f and ϕ are changed simultaneously. Adapted with permission from *Opt. Lett.* **25**, 16, pp. 1166. Kourogi et al. (2000).

was observed with a scanning Fabry-Perot cavity and an optical spectrum analyzer. The linewidth of laser is estimated to be less than 1 MHz by beat note signal. Fig. 6.15(a) shows the wavelength variation when f is tuned without a delay line. The discontinuities of the tuning wavelength are due to the Fabry-Perot modes of the internal cavity of the semiconductor laser itself.

Fig. 6.15(b) shows an enlarged version of Fig. 6.15(a). It is evident that the laser experiences a stairway-shaped mode hopping, owing to the Fabry-Perot mode of external cavity, where the mode-hop frequency is equal to the free spectral range of the Fabry-Perot modes ~1 GHz of the external cavity. This behavior agrees with that observed in Fig. 6.14(a). From Fig. 6.15 it is estimated that $\langle df_{laser}/df \rangle = 3 \times 10^4$, and thus the optimum value of τ is about 15 μs, Fig. 6.15(c) and (d) show the behavior of the

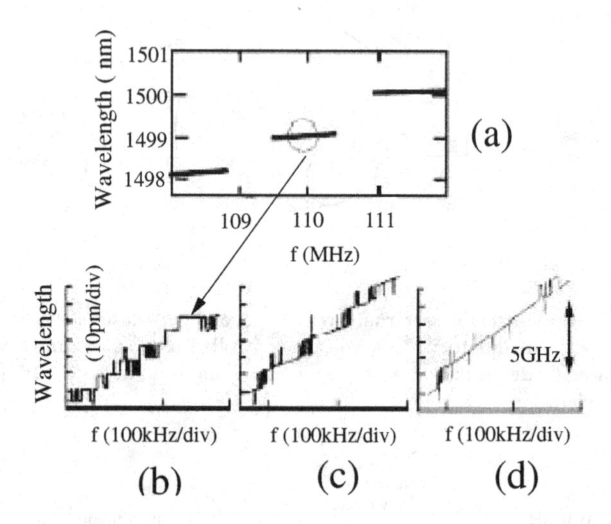

Fig. 6.15 Behavior of the wavelength of laser as a result of frequency tuning: (a), (b) $\tau = 0$; (c) $\tau = 7.5$ μs; (d) $\tau = 15$ μs. Adapted with permission from *Opt. Lett.* **25**, 16, pp. 1166. Kourogi et al. (2000).

wavelength when $\tau = 7.5$ μs and 15 μs, respectively. To achieve such a large delay one utilizes an optical delay line. It is found that wavelength behaves in a different way when f is tuned with a different delay line. When $\tau = 7.5$ μs, a stairway-shaped mode hopping can be seen. It can be seen clearly that laser is being tuned by f and ϕ being changed simultaneously. When $\tau = 15$ μs, the stairway-shaped mode hop changes to continuous tuning for several gigahertz except for some random mode hopping. This approach lends itself to fully electronic control of external cavity diode lasers.

6.2.2 *Frequency chirping by electro-optic modulator*

Integrated optic filter based on narrow-band forward grating-coupling between TE and TM modes in $LiNbO_3$ has been demonstrated [Alferness and Buhl (1982)]. With the addition of electro-optic phase shifters, these devices have the potential to provide both mode selection and mode shifting for continuous access tuning. Tuning ranges of 700 nm and linewidth of ≤ 60 kHz have been obtained with such devices [Heismann *et. al.* (1987)]. External cavity diode lasers have proven to be a very powerful tool for a variety of applications. The ECDLs ability to tune to particular transitions makes then invaluable for studies in atomic physics [Wieman and Hollberg (1991)]. However, new applications are arising requiring ECDLs capable of producing linear frequency chirps. Frequency chirped lasers have proven to be very useful for optical coherent transients processing [Merkel and Babbitt (1996); Kroll and Elman (1993)], memory [Lin *et. al.* (1995); Bai *et. al.* (1986)], true time delay and coherent laser radar [Karlsson and Olsson (1999)]. Recently, the methods of producing the frequency chirp for laser radar have been demonstrated.

The first method involves the modulation of the drive current of a distributed feedback laser diode to produce the frequency chirp [Karlsson and Olsson (1999)]. The response of the laser to a change in current is not a linear function so that a complicated modulation function is needed to produce a linear frequency chirp. The second method used to produce a frequency of diode laser is to provide frequency shifted feedback to the laser diode using an intra-cavity acousto-optic modulator [Nakamura *et. al.* (1998); Nakamura *et. al.* (1998a); Kowalski *et. al.* (1998); Troger *et. al.* (1999)]. The output of a laser diode is collimated and sent through the AOM, the first order beam provides the frequency shift feedback. The slope of the frequency chirp is $\gamma = \nu_f/\tau_r$, where ν_f is the frequency shift per round trip time while the duration of the frequency chirp is $1/\nu_f$, τ_r is round trip time.

The third method has been developed by using an intra-cavity electro-optic crystal to chirp an external cavity laser [Tang *et. al.* (1977); Goedgebuer *et. al.* (1992); Wacogne *et. al.* (1993); Andrews (1991); Boggs *et. al.* (1998); Mnager *et. al.* (2000)]. A resonance condition is set by $l_{cav} = q\lambda/2$ so that an integer number of half wavelengths fits into the optical path length l between the back facet of the laser diode and the grating that forms the external cavity. By applying a voltage to the electro-optic crystal, the index of refraction changes and thus optical length changes by $\Delta l = \Delta n_e(V)l$, where $\Delta n_e(V) = -(n_e^3/2)r_{33}(V/d)$ is the change in the index of refraction, n_e is the extraordinary index of refraction of the EO crystal with no applied voltage, r_{33} is the electro-optic coefficient of the EO crystal, V is applied voltage, and d is the spacing between the electrodes. The chirped ECDLs wavelength thus changes to match this new resonant condition by $l_{cav} = q\lambda/2$.

One can write an expression for the change in frequency as a function of the applied voltage

$$\Delta\nu = \frac{\nu n_e^3 l r_{33}}{2 l_{cav} d} V = R_{eo} V, \tag{6.3}$$

where l is the length of EO crystal, R_{eo} is a constant that describes the response of the EO crystal to the applied voltage. The change in frequency of the laser is linearly proportional to the applied voltage. The time of the frequency chirp of a chirp ECDLs with an intra-cavity EO crystal can be programmed by controlling the voltage of the electro-optic crystal. Thus flexibility in programming the timing of the frequency chirp is a major advantage over the frequency shifted feedback laser.

A 1540 nm external cavity laser diode with an intracavity electro-optic wavelength tuner has been reported [Wacogne *et. al.* (1994)], coarse- and fine-wavelength filter has been achieved with a single lithium niobate wavelength tuner. One section of the crystal is used as a tunable-wavelength filter to tune the wavelength by mode hoping. The other operates as a phase modulator to change the optical length of the cavity and to tune the wavelength continuously.

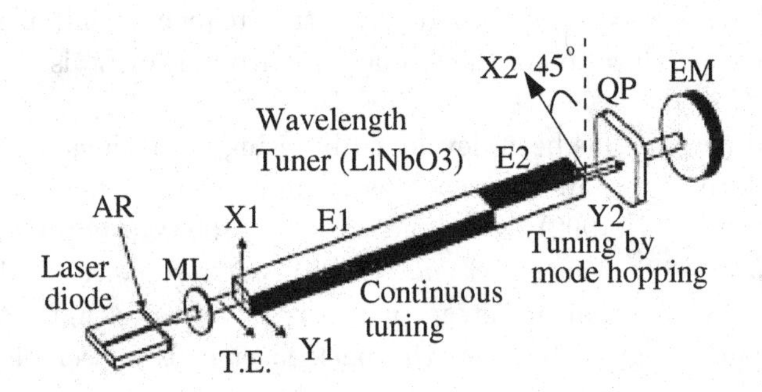

Fig. 6.16 Wavelength-tunable laser diode: ML, microlens. QP, quartz plate, EM. external mirror. Adapted with permission from *Opt. Lett.* **19**, 17, pp. 1334-1336. Wacogne et al. (1994).

The experimental setup is shown schematically in Fig. 6.16, it consists of a laser diode, a collimating microlens, an electro-optic wavelength tuner, a birefringent quartz plate, and an external mirror. A diode laser is a 1540 nm GaAsInP/InP with one internal facet AR coated. The wavelength tuner is a z-propagating bulk lithium niobate crystal with two pairs of electrodes oriented at 90 ° one to each other. The crystal length is 45 mm along the z direction, and its cross section is 2 mm × 2 mm. The length of long electrodes is d_1=35 mm. These electrodes induce an electrical field along the y_1 axis that is parallel to the TE polarization of the laser chip. The length of the short electrodes (d_2=10 mm) induces an electric field along the x_1 axis of the crystal. The fast x_2 and slow y_2 axes of the quartz plate are oriented at 45° to TE. The optical path difference induced by natural birefringence of the plate between its slow and fast axes is $D_0 = 57\ \mu m$, the optical cavity length is $L = 16\ cm$.

Tuning with mode hopping is obtained with the short electrode section of the crystal, which is in series with the birefringent quartz plate. When a voltage is applied to this short electrode section, the resulting voltage-dependent optical path difference between two x_2 and y_2 axes is

$$D(V_2) = d_2 n_o^3 r_{12}(V_2/w), \tag{6.4}$$

where $n_o = 2.2116$ is the ordinary refractive index of lithium niobate at 1540 nm wavelength, $r_{12} = 3.4 \times 10^{-12} m/V$ is the electro-optic coefficient and w=2 mm is the aperture width of the crystal. For one round trip of laser in the laser cavity, the spectral transmission of such a filter is given by

$$T(\nu) = \cos^2[\frac{2\pi\nu}{c}(D_0 + D(V_2))], \tag{6.5}$$

transmission peak occur at frequencies $\nu = qc/2[D_o + D(V_2)]$, where q is an integer.

Laser emission occurs at the wavelength of the maximum gain. Single-mode operation is achieved as a single peak is inside the gain curve to avoid more mode competition. One then obtains wavelength tuning with mode hopping by varying the peak position in the gain curve through voltage V_2, then the tuning rate with mode hopping is given by

$$d\nu/dV_2 = 2d_2 c n_o^3 r_{12}/qw\lambda_o^2 = 630\ MHz/V, \tag{6.6}$$

where q=74 is the transmission peak order for $d_2 = 10\ mm$ at λ_o=1540 nm. Continuous wavelength tuning is obtained by phase modulation of the intracavity wave by means of the long-electrode section. When a voltage V_1 is applied to the crystal, whose axes are not rotated as described above, but the n_{y1} refractive index is electrically changed by the Pockels effect:

$$n_{y1} = n_0 + \frac{1}{2} n_0^3 r_{12} V_1/w, \tag{6.7}$$

considering the variation of the refractive index n_{y1} with applied voltage V_1, the fine-voltage-tuning rate can easily be written as

$$d\nu/dV_1 = cn_o^3 r_{12} d_1/(2w\lambda L) = 400\ kHz/V, \tag{6.8}$$

where $d_1 = 35\ mm$.

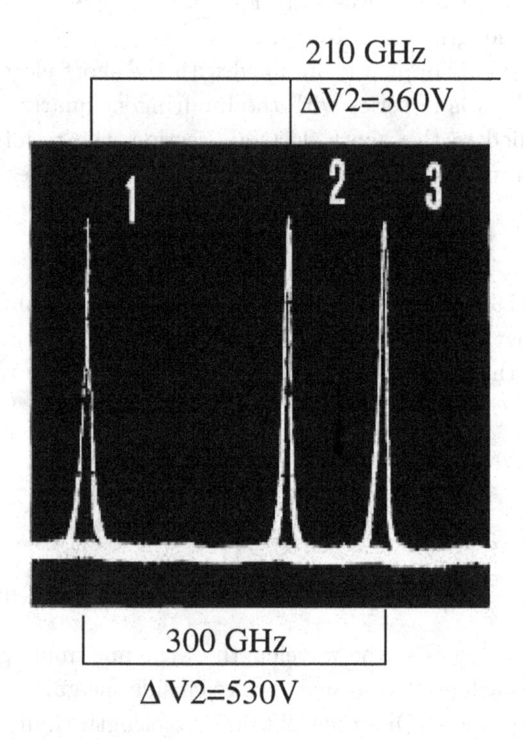

Fig. 6.17 Wavelength tuning by mode hopping (multiple exposure). Adapted with permission from *Opt. Lett.* **19**, 17, pp. 1334-1336. Wacogne et al. (1994).

Figure 6.17 is multiple-exposure photography of wavelength tuning by mode hopping obtained with a Fabry-Perot spectrum analyzer when voltage V_2 is applied to the short-electrode section. The measured tuning rate is 580 MHz, The largest tuning range is 700 GHz. Figure 6.18 shows an example of continuous tuning obtained by applying the voltage V_1 across the longer section of the crystal. The maximum tuning range was 770 MHz, The corresponding driving voltage is 400 V, which gives a continuous tuning rate of 480 kHz/V.

The performance of a chirped ECDLs built in the Littrow configuration

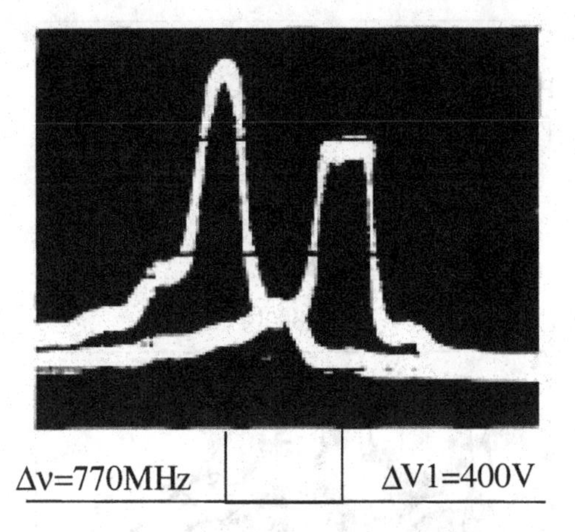

Fig. 6.18 Continuous tuning over 770 MHz, i.e., 82% of the FSR of the laser cavity (multiple exposure). Adapted with permission from *Opt. Lett.* **19**, 17, pp. 1334-1336. Wacogne et al. (1994).

with an intra-cavity electro-optic crystal has been studied [Rapasky *et. al.* (2002)]. The chirped ECDLs has a center wavelength of 793 nm with a 20 nm tuning range by mechanical rotation of the feedback grating and rapid tuning is achieved electronically by voltage control of the electro-optic crystal. The EO tuning response of the laser is 2.01 MHz/V, and linear frequency chirp of 800 MHz ranging in duration from 3 μs to 337 μs has been demonstrated. The maximum electro-optic tuning is set by the external cavity mode spacing of 2.4 GHz.

A schematic of the chirped ECDLs is shown in Fig. 6.19(a) and the actual laser is shown in Fig. 6.19(b). Light from a laser diode with a nominal wavelength of λ=795 nm is collimated with a lens of focal length f=4.50 mm and a numerical aperture of NA=0.55, the lens has a broadband anti-reflection coating. After the collimating lens, the light is incident on an EO crystal. The crystal is a x-cut lithium tantalite crystal that is $5 \times 5 \times 10\ mm^3$. Two opposing $5 \times 10\ mm^2$ sides are coated with a golden plating and are used for the high voltage electrical connection. After the EO crystal, the light is incident on an 1800 line/mm holographic grating at an incident angle of 45°. The first order reflection from the grating which is used to provide the optical feedback from the grating is incident on a roof prism. The grating and roof prism form a retroreflector so that as the laser is tuned, the output does not change its pointing direction.

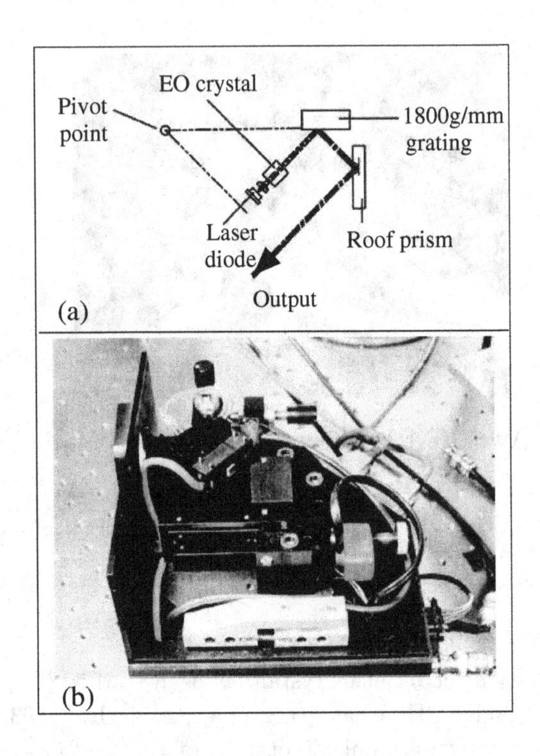

Fig. 6.19 (a) Schematic of the CECDL design. The EO crystal provides a method for chirping the laser, the pointing direction of the laser output is key constant as the laser is tuned because of the retroreflector formed by the grating and roof prism. (b) Picture of the completed chirped ECDL. Adapted with permission from *Rev. Sci. Instrum.* **73**, 9, pp. 3155. Rapasky et al. (2002).

A commercial laser diode controller is used to set the current and control the temperature of the chirped ECDLs. The voltage applied to EO crystal can be from high voltage supply or an amplified signal from a programmable ramp driver. The mechanical tuning of the chirped ECDLs is controlled by a single channel piezo drive. Figure 6.20 is a plot of the laser frequency in GHz as a function of the applied voltage to the electro-optic crystal. The electro-optic tuning of the laser via the electro-optic crystal shows a tuning range of 1.2 GHz without mode hop, which is less than the 2.4 GHz mode spacing of the cavity. The open circle represent the measured values of the laser frequency as a function of the applied voltage while the solid line is a linear fit to the data. The slope of the linear fit of $R_{eo} = 2.01 \pm 0.02$ MHz/V, compared to theoretical calculation of R_{eo}=1.99 MHz/V, agrees well with theory.

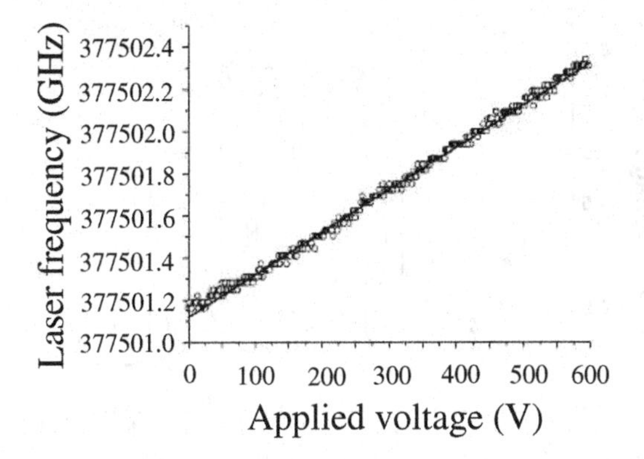

Fig. 6.20 Plot of the laser frequency as a function of the voltage applied to the EO crystal. This measurement was made with the SRS voltage supply and wavemeter and thus represents the dc detuning response due to the EO crystal. The solid line is a linear fit to the data and has a slope of ±0.02 MHz/V. Adapted with permission from *Rev. Sci. Instrum.* **73**, 9, pp. 3157. Rapasky et al. (2002).

The expected response of the chirped ECDL $\Delta\nu(t)$ can be expressed in terms of the measured response of the electro-optic crystal R_{eo} as follows:

$$\Delta\nu(t) = V(t)G_{HV}R_{EO}, \tag{6.9}$$

where V(t) is the measured voltage output from the ramp generator, G_{HV}=200 is the gain of the high voltage amplifier. The chirped frequency response of the chirped ECDL is measured using the scanning FP interferometer. When the chirped ECDL is not chirping, the transmitted intensity of the scanning interferometer has a Lorentzian line shape as a function of time.

The transmitted intensity as a function of time will deviate from the Lorentzian line shape and this deviation is directly related to the frequency chirped response. The deviation from the Lorentzian line shape can then be used to find the frequency chirped response of the chirped ECDL. The solid line in Fig. 6.21 shows the transmitted intensity as a function of time for chirped ECDL with a linear voltage ramp of 400 V repeated applied to the EO crystal in 337 μs while the dashed line is the Lorentzian envelope. The deviation from the Lorentzian is evident in this figure.

Fig. 6.22 is a plot of the frequency chirped response of the chirped ECDL as a function of time. The circles represent the measured value of the frequency response measured by looking at the deviations from the

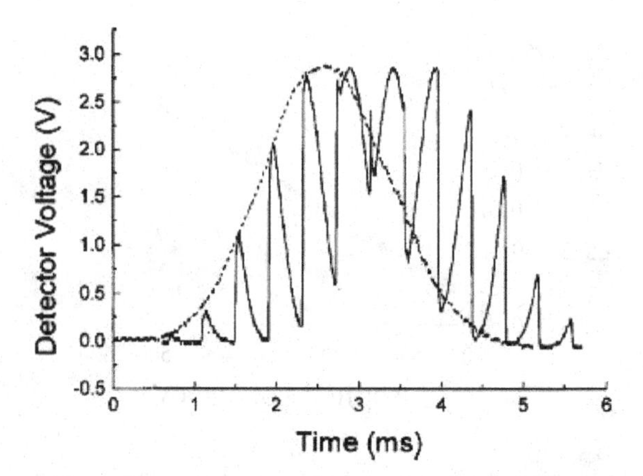

Fig. 6.21 Plot of the transmitted power of the scanning FP interferometer as a function of time. The solid line is measured when the CECDL is not chirping and thus has a Lorentzian line shape. The deviation from a Lorentzian line shape is evident when the CECDL is chirping and this deviation is used to find the frequency chirped response as a function of time. Adapted with permission from *Rev. Sci. Instrum.* **73**, 9, pp. 3158. Rapasky et al. (2002).

Lorentzian line shape as shown in Fig. 6.21. The solid line is the expected frequency response as calculated from Eq. (6.9). Good agreement is seen between the measured and expected responses.

Figure 6.23 is a plot of the frequency chirped response of chirped ECDL as a function of time for linear voltage ramps applied to the electro-optic crystal. For this plot, voltage ramp duration of 3, 13, 28, 93, 195 and 337 μs were used. The good agreement between the measured and expressed frequency response indicates that an intra-cavity electro-optic crystal is a good way to produce linear frequency chirps on the microsecond time scale.

A novel tuning concept has been proposed and demonstrated[Levin (2002)], where frequency tuning range is increased by making the electro-optic crystal thinner. The laser has the chirp rate of 1.5 GHz/μs and good frequency stability. The change in cavity length that is imposed by the electro-optic crystal is inversely proportional to the crystal thickness. If the crystal thickness changes across the beam perpendicular to the propagation direction, one can achieve the difference in the cavity elongation that is required for the mode-hop-free tuning. The proper shape of the crystal can be derived as follows: If θ is the angle of incidence on the grating, y is

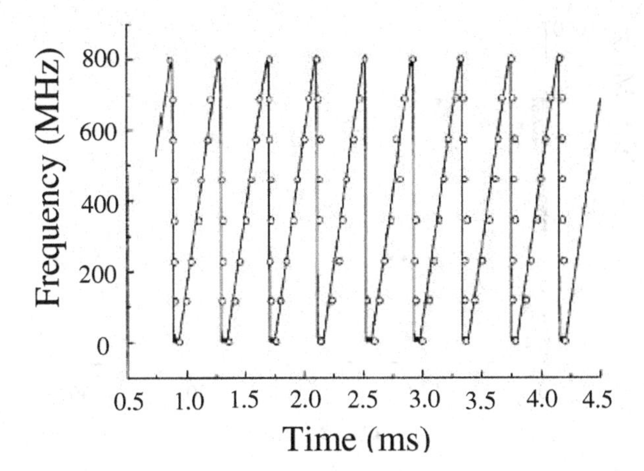

Fig. 6.22 Plot of the laser frequency chirp as a function of time. The circles represent measured values of the frequency chirp by using the data in Fig. 6.21 while the solid line is the expected frequency chirped response calculated from Eq. (6.9). Adapted with permission from *Rev. Sci. Instrum.* **73**, 9, pp. 3158. Rapasky et al. (2002).

a coordinate as defined in Fig. 6.24, then the optical length L(y) is given by L(y)=L(0)+ytan θ. If the height of the crystal h(y) varies slowly, then the change in the cavity length with applied voltage as a function of y is given by

$$\Delta L(y) = -\frac{a}{2} n_z^3 r_{33} \frac{V}{h(y)}, \tag{6.10}$$

where a is the length of the crystal, $n_z = n_e$ is the extraordinary refractive index, r_{33} is the relevant element of the electro-optical tensor, and V is applied voltage.

The conclusion of mode-hop-free tuning is that the elongation of the cavity should be proportional to the cavity length for all y. Therefore, it requires that $\Delta L(y) = CL(y)$, where C is a constant. That means

$$h(y) = -\frac{\frac{a}{2} n_z^3 r_{33} V}{[L(0) + y \tan\theta] C} = h(0) \frac{1}{1 + \frac{\tan\theta}{L(0)} y}, \tag{6.11}$$

it is easier to design a crystal with flat surfaces. If the beam width is considerably smaller than the cavity length, one can have

$$h(y) \approx h(0)[1 - \frac{\tan\theta}{L(0)} y]. \tag{6.12}$$

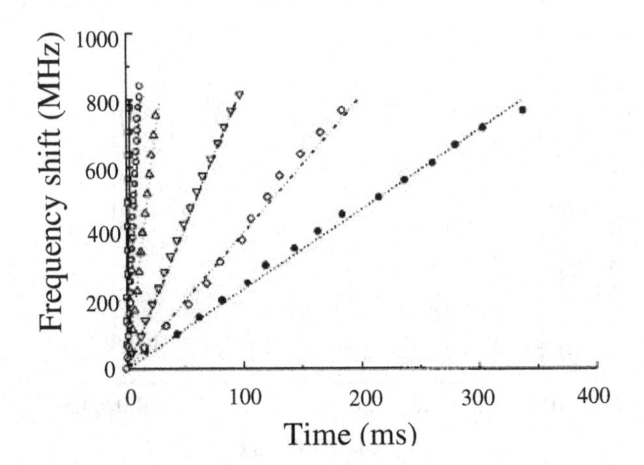

Fig. 6.23 Frequency chirped response of the CECDL for chirp duration ranging from 3 to 3.337 μs. The solid line represents the expected frequency chirp while the symbols represent the measured values. Adapted with permission from *Rev. Sci. Instrum.* **73**, 9, pp. 3158. Rapasky et al. (2002).

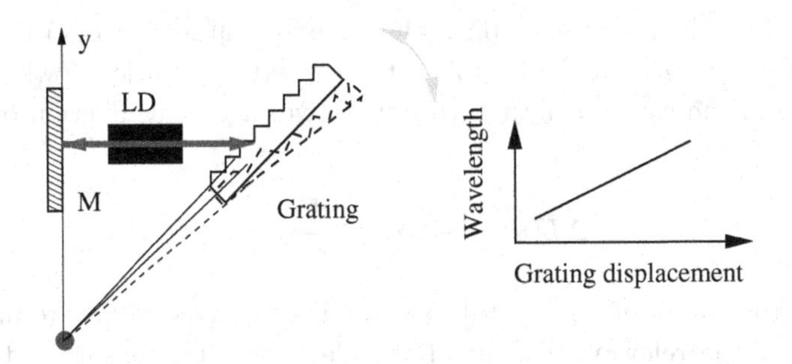

Fig. 6.24 Schematic diagram of moving the grating and consequently tuning the wavelength. Adapted with permission from *Opt. Lett.* **27**, 4, pp. 237-239. Levin (2002).

To get as high a frequency voltage tuning ratio as possible, it is desirable to get thin crystal. Stable single-mode operation is ensured by use of large value of θ. The schematic of setup is shown in Fig. 6.25(a). The 50 GHz continuous single-mode scanning range is shown in Fig. 6.25(b). The dots on the line are recorded measurement points, but the change is monitored continuously. The mode-hop-free is monitored by observing the smooth movement of the fringes from a scanning Fabry-Peort interferometer. Fig. 6.26(a) shows a recording of a 400 MHz scan by applying a

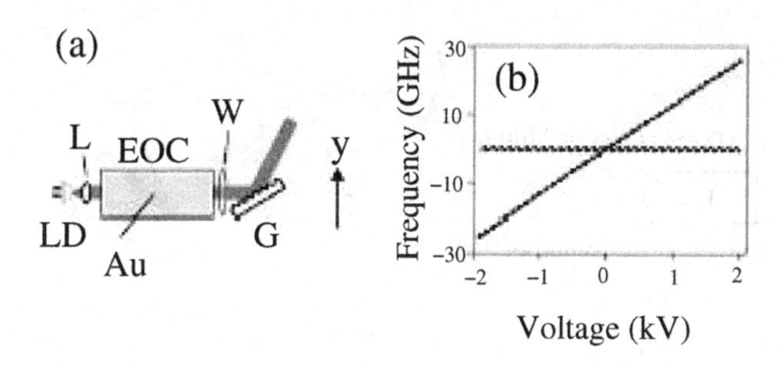

Fig. 6.25 (a) Schematic of laser setup. LD: laser diode; L: lens; Au: Au-plate; EOC: electro-optic crystal; W: half-wave plate; G: grating. (b) The line with dots is a recording of a 50 GHz scan. The sawtooth line shows how the tuning would have appeared with a nonangled crystal, limiting the scanning to range to 1.3 GHz. Adapted with permission from *Opt. Lett.* **27**, 4, pp. 237-239. Levin (2002).

high voltage across the crystal. A scan with a maximum scanning speed of 1.5 GHz/μs is also shown in Fig. 6.26(b). The fast scanning capabilities

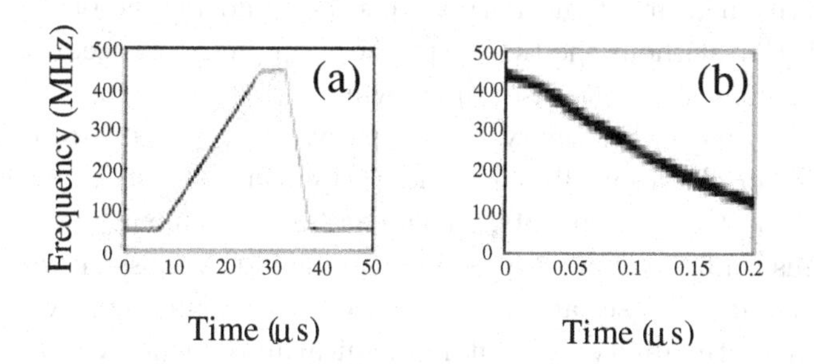

Fig. 6.26 Recording of a 400 MHz chirp (a) and of fast scan (b). Adapted with permission from *Opt. Lett.* **27**, 4, pp. 237-239. Levin (2002).

and short-term stability of the laser are analyzed by means of heterodyne mixing the laser with the other laser with linewidth of 100 kHz.

6.2.3 *Wavelength tunability by liquid crystal display*

A new concept of an electronically tunable external-cavity diode laser with simultaneous feedback and intracavity spatial separation of the laser's spectral components has been demonstrated by use of liquid crystal spatial light modulator [Struckmeier *et. al.* (1999)] . As depicted in Fig. 6.27, the external cavity consists of an AR coated commercial 670 nm laser diode, a

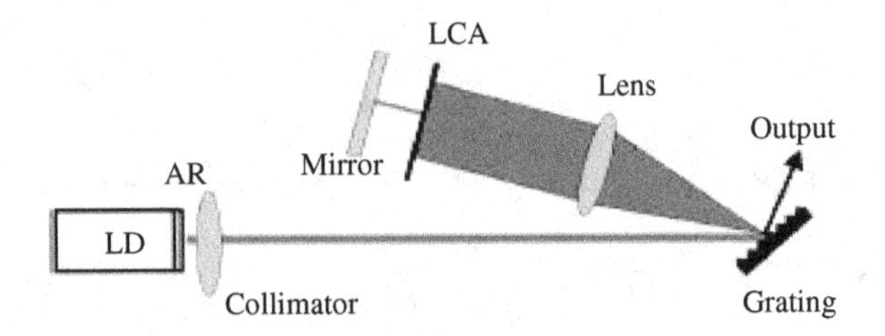

Fig. 6.27 Schematic diagram of the electronically tunable external-cavity diode laser. Adapted with permission from *Opt. Lett.* **24**, 22, pp. 1573-1575. Struckmeier et al. (1999).

collimator, and diffraction grating, a liquid-crystal array (LCA), and a high-reflection end mirror. The output beam of the laser diode is collimated and sent into the diffraction grating. The grating is placed such that its first diffraction order is directed toward the lens and that the distance between grating and lens equals the focal length of the lens. The high-reflection mirror is placed in the other focal plane of the lens, and the LCA is located directly in front of the mirror. The basic advantage of this cavity geometry is that without the LCA it provides simultaneous feedback for all spectral components of the laser diode while the spectral components are spatially separated in the cavity. This separation enables one to introduce the LCA as an electronically controllable aperture to select feedback for the various spectral components, which can be independently switched on and off. This configuration permits both tuning of the emission wavelength and operation of the laser at various wavelengths simultaneously.

Wavelength tunability without mechanical movement over a range of 10 nm between 665 nm and 676 nm is shown in Fig. 6.28. The linewidth of the laser was measured with Fabry-Perot interferometer to be smaller than 30 MHz with a single-mode suppression of better than 10 dB. The possibility of multi-color synchronous operation is one of the great advantage of the electronically tunable external cavity laser diode. Two-color operation are demonstrated with a 670 nm laser diode as shown in Fig. 6.29. Dual emission is obtained at various spectral positions and with variable spacing between the emission modes. Using an intracavity Fourier transformation allows one to demonstrate the multi-color operation, gain extension and intracavity second harmonic generation in place of LCA with LCD or digital mirror devices as shown by the figure similar to Fig. 6.27. The new design can be applied to different lasers as shown on the examples of a laser diode and a Thulium-doped fiber laser in the mid-infrared at wavelength

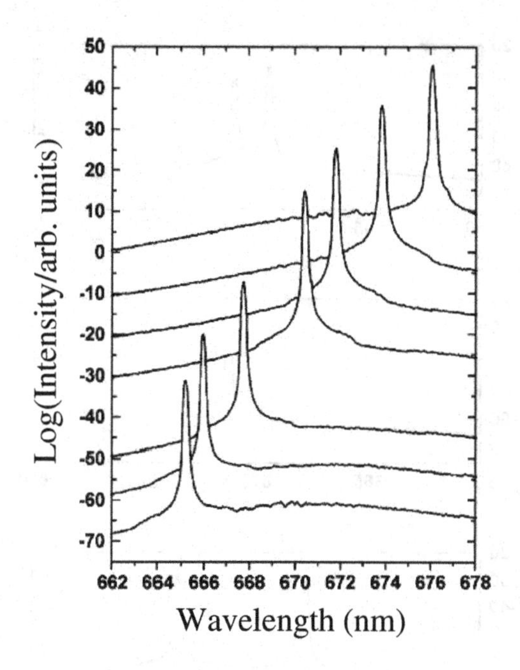

Fig. 6.28 Spectra of electronically tunable external-cavity diode laser emission for several settings of the LCA. The traces have been vertically offset. Adapted with permission from *Opt. Lett.* **24**, 22, pp. 1573-1575. Struckmeier et al. (1999).

2300 nm. In addition, tunable emission in the 490 nm range is obtained by use of intracavity SHG with one single control parameter [Breede *et. al.* (2002)].

6.3 Miscellaneous external cavity diode lasers

In this section, miscellaneous external cavity diode laser systems are described, including blue-violet diode lasers which has successfully developed recently, high power diode laser as an important class of applications requiring high power either in pulsed or continuous-wave operation in a single-frequency mode, and wide tunability achieved by latest quantum dots lasers in an external cavity.

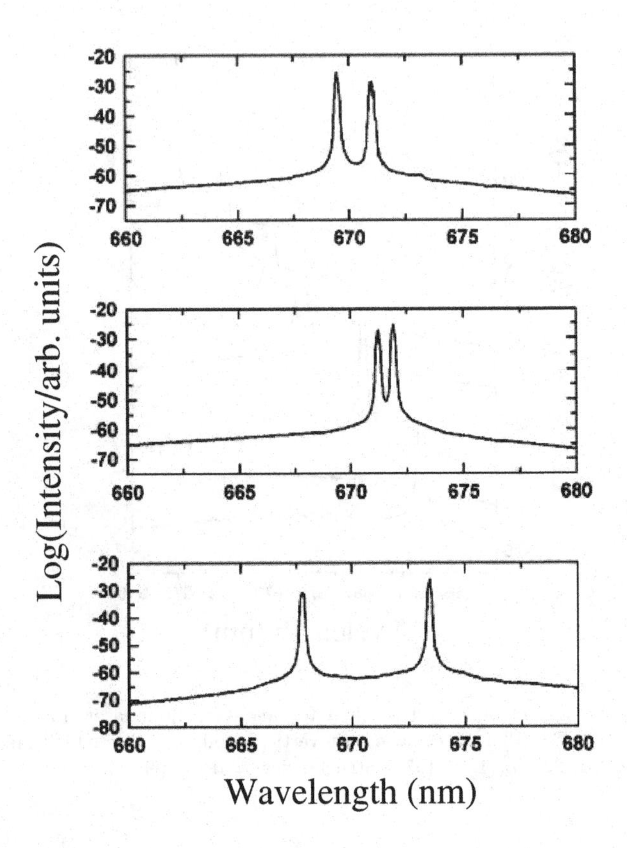

Fig. 6.29 Three emission spectra of electronically tunable external-cavity diode laser with dual-color operation. Adapted with permission from *Opt. Lett.* **24**, 22, pp. 1573-1575. Struckmeier et al. (1999).

6.3.1 *Blue-violet external cavity diode lasers*

The rapid development of gallium nitride (GaN) diode laser has led to the recent commercial availability of CW blue-violet laser diode with its wavelength range from 370 nm~430 nm. The main market of these laser is not only for optical data storage, but also for scientific applications. It is a good alternate to frequency-doubled dye or Ti: sapphire lasers.

Conroy et al. [Conroy *et. al.* (2000)] reported a detail characteristics of a GaN diode laser operating in a Littrow configuration in an external cavity around 392 nm, giving a tunable, single frequency output power in excess of 3.5 mW. The laser can be tuned smoothly over 6 GHz and 2.7 nm discontinuously with a upper limit linewidth of 5 MHz. Figure 6.30

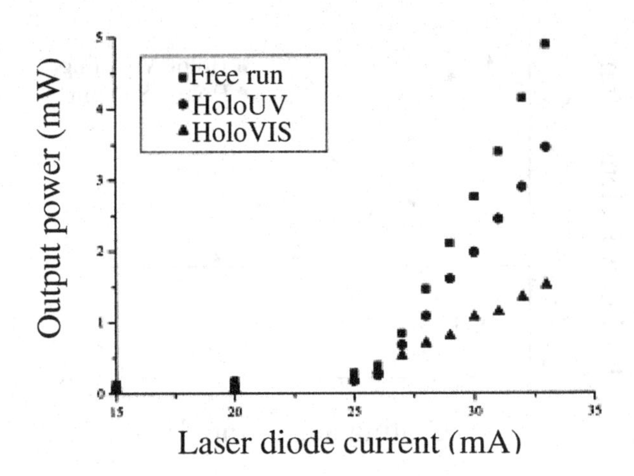

Fig. 6.30 Output power of free-running and external-cavity lasers as a function of laser current. Adapted with permission from *Opt. Commun.* **175**, pp. 186. Conroy et al. (2000).

shows the standard output characteristics of the Nichia Corporation NLHV 500 with nominal cw output power of 5 mW at 393 nm in the case of free running and external cavity in Littrow configuration. As indicated in Fig. 6.30, it can be seen that the holographic UV (HoloUV) grating with lower diffraction efficiency (30%) gives a larger slope efficiency of 46%, with a maximum of 3.5 mW of single frequency output power. The more efficient holographic visible (HoloVIS) (50%) grating shows a measurably reduced threshold of 26.4 mA but with a ower slope efficiency of 17%, thereby limiting the maximum available single frequency power to 1.5 mW.

The tuning range of the external cavity is illustrated in Fig. 6.31 as the angle and position of the grating is changed manually. The tuning range are 1.3 nm and 2.7 nm for HoloUV and HoloVIS, respectively. To compare this tuning behavior, Fig. 6.32 shows the tuning behavior from 635 nm and 670 nm laser diodes with same configuration. Both lasers are anti-reflection coated with a reflectivity of less than 10^{-5}, one finds the discontinuous tuning range of up to 10 nm for these two diode lasers. The limited discontinuous tuning of GaN laser diodes in an external cavity is thus a disadvantage. The output could be continuously tuned by means of a lower voltage piezoelectrical transducer to change the grating angle and its distance from the diode laser. The tuning of 6 GHz without mode-hopping has been achieved by a voltage of 10 V peak to peak. A linewidth of 5 MHz

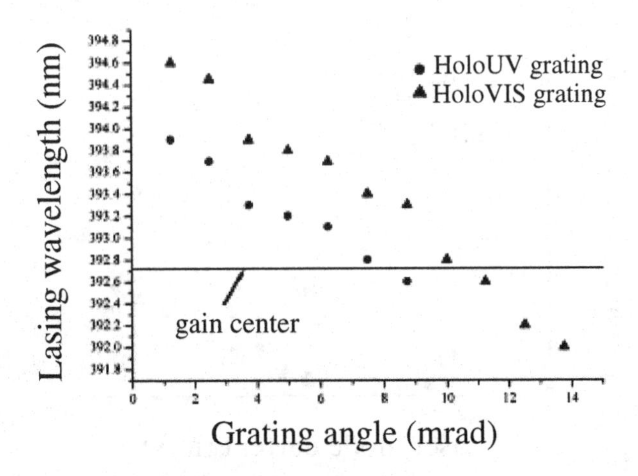

Fig. 6.31 Discontinuous wavelength tuning of the violet external-cavity diode laser using both the HoloUV and HoloVIS gratings. The gain center of the free-running diode was at 392.7 nm. Adapted with permission from *Opt. Commun.* **175**, pp. 187. Conroy et al. (2000).

single mode by scanning etalon is shown in Fig. 6.33. The application of this type of laser in high-resolution spectroscopy and atom/ion trapping will be given in chapter 8.

A new ECDLs system with a filter cavity have been developed recently [Hayasaka (2002)], which reflects the most part of the background. The experimental setup with a violet diode laser at 397 nm is shown in Fig. 6.34, An output power of 13.2 mW from ECDLs is coupled to a ring cavity with a measured finesse of 400, approximate 74% of input beam power is coupled into the cavity, and a output power of 4.1 mW is transmitted through the output mirror. The locking of the ECDL frequency to the cavity is done by resonant optical feedback. The spectra of the violet laser observed with an optical spectrum analyzer is shown in Fig. 6.35(a). Despite the fact that many small longitudinal modes were observed in the ECDLs (Fig. 6.35(b)), they are completely eliminated by the filter cavity (Fig. 6.35(c)).

The spectroscopic properties of one of the first samples of blue-emitting diode based on GaN was characterized by Leinen et al. [Leinen *et. al.* (2000)]. With Such a laser diode operated in a standard external cavity arrangement, a mod-hop free tuning range of more than 20 GHz and a linewidth of 10 MHz was found. The application of this violet lasers to high-resolution spectroscopy will be introduced in chapter 8 as well.

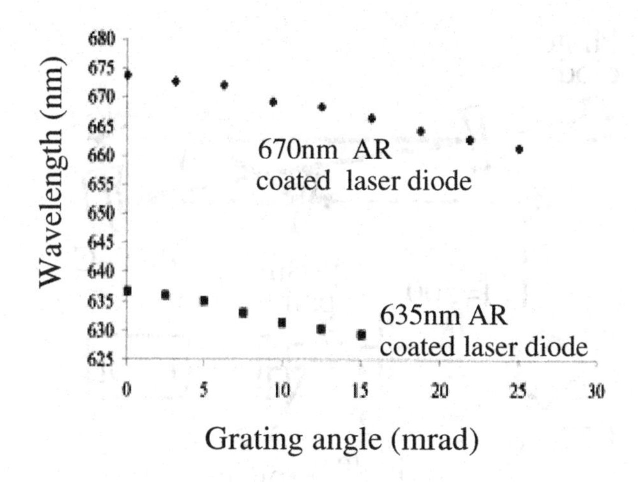

Fig. 6.32 Discontinuous wavelength tuning of both 670 nm and 635 nm laser diodes in the compact external-cavity geometry. The tuning range is much larger than these for the violet ECDL. Adapted with permission from *Opt. Commun.* **175**, pp. 187. Conroy et al. (2000).

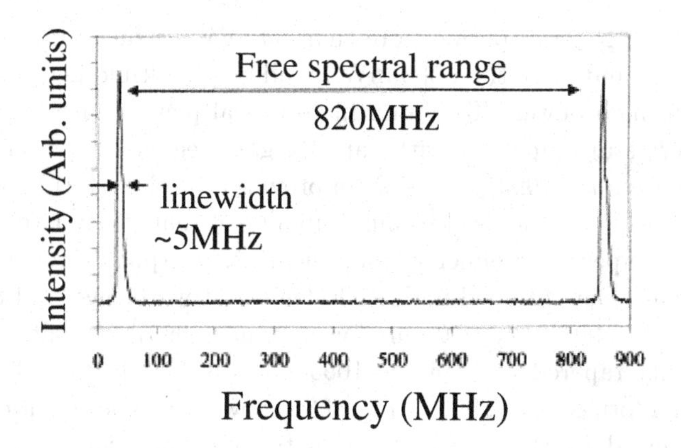

Fig. 6.33 A measurement of the linewidth of the violet ECDL using a Fabry-Perot etalon with FSR of 820 MHz and a finesse of 200. Adapted with permission from *Opt. Commun.* **175**, pp. 187. Conroy et al. (2000).

6.3.2 *High power external cavity diode lasers*

The distinct difference between lower power and high power semiconductor lasers is not clearly defined. It may depend on the type of laser and its

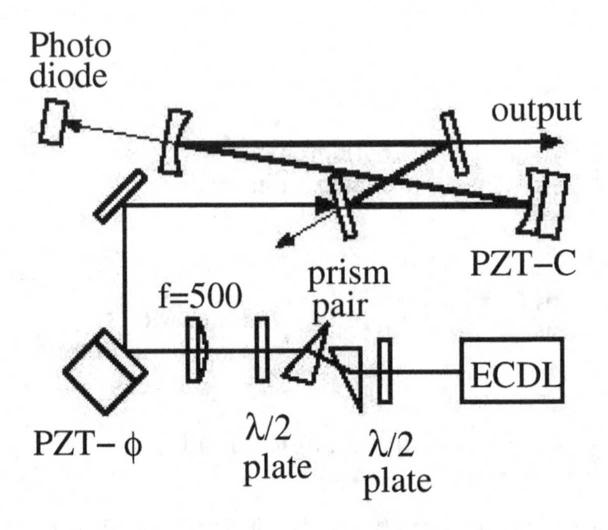

Fig. 6.34 Schematic of experimental setup. Adapted with permission from *XVIII International Conference on Atomic Physics,* Cambridge, Massachusetts, USA, pp. 146. Hayasaka and Uetake (2002).

applications. In general, power more than 50 mW for single-mode, single frequency laser and more than 500 mW broad area multimode diode lasers can be called high power. To obtain high optical powers with high beam quality, lasers and amplifiers with tapered gain region have been developed [Kintzer *et. al.* (1993)], the system provides wavelength tunability. A diffraction grating can be used to build up an external cavity laser system including the tapered amplifier as gain element. Output power in excess of 1 W cw has been obtained with such devices at wavelengths of 850 nm [Mehuys *et. al.* (1997)] and 970 nm [Jones *et. al.* (1995)]. A grating tuned external cavity tapered laser in the 1055 nm wavelength range has been reported [Morgott *et. al.* (1998)]. Fig. 6.36 shows the experimental setup, the light emitted by the narrow facet of tapered amplifier is collimated by an aspheric lens (f=6.5 mm NA=0.62) onto a diffraction grating with 1200 line/mm, which is mounted in the Littrow configuration. The grating is oriented such that the grating lines are parallel to the active region to disperse the spectrum perpendicular to the epilayers.

The configuration has two advantages: firstly, the diameter of the collimated beam is larger in the vertical direction and therefore more grating lines are illuminated in this configuration, that leads to a larger wavelength dispersion. Secondly, the only 1 μm thick vertical waveguide of the am-

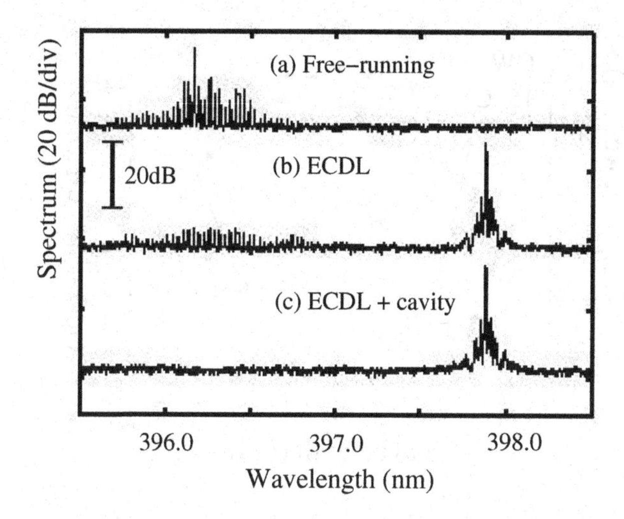

Fig. 6.35 Spectra of the violet diode laser. Adapted with permission from *XVIII International Conference on Atomic Physics,* Cambridge, Massachusetts, USA, pp. 146. Hayasaka and Uetake (2002).

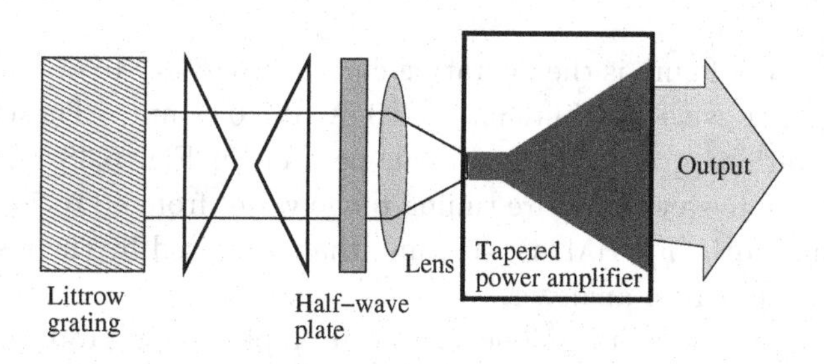

Fig. 6.36 Schematic experimental setup of grating tuned external cavity laser. Adapted with permission from *Electron. Lett.* **34**, 6, pp. 558. Morgott et al. (1998).

plifier acts as an entrance slit of a monochromator to capture the lasing wavelength. To increase the reflectivity of the grating into the first order by 13%, a half-wave plate is inserted into the external cavity which rotates the direction of polarization perpendicular to the grating lines [Lotem *et. al.* (1992)].

The tuning curve of the external cavity laser for CW operating is shown in Fig. 6.37, An output power of more than 1 W is obtained over the wide tuning range of 55 nm, from 1030 nm to 1085 nm at a current of 4 A.

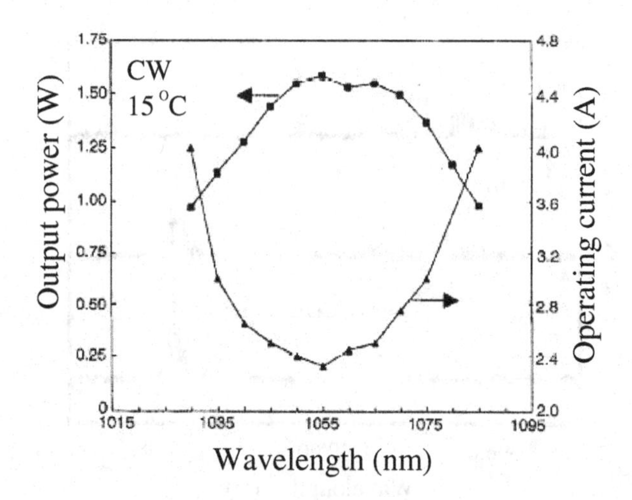

Fig. 6.37 Tuning curves of external cavity laser. Adapted with permission from *Electron. Lett.* **34**, 6, pp. 558. Morgott et al. (1998).

Also shown in the figure is the operating current necessary to obtain 1.0 W of output power with a minimum of 2.3 A at 1055 nm. The emission spectra for various grating position can be seen in Fig. 6.38. The side mode suppression over the entire tuning range varies from 40 to 50 dB, the measured linewidth (FWHM) is ~0.1 nm, that is limited by the resolution of the optical spectrum analyzer.

An injection-locking technique has been employed to produce single-frequency tunable output from a high-power A1GaAs laser diode array [Tsuchida (1994)]. A narrow-linewidth external cavity laser is used as a master oscillator. With an injected power of 15 mW, a total output power of 1.0 W was obtained from the laser diode array with a spectral linewidth of less than 38 kHz, a wavelength tuning range of 13 nm, and a 0.67 ° wide single-lobed far-field beam. More recently, Bayram and Chupp[Bayram and Chupp (2002)] have developed the commercial 2 W laser diode array in standard Littman-Metcalf configuration, this external-cavity laser diode array system is operating on a dominant single longitudinal mode of narrow linewidth of 120 MHz and power of 1 W or more. This improvement enables applications such as spectroscopy, laser cooling and trapping with relatively inexpensive high-power laser diode arrays.

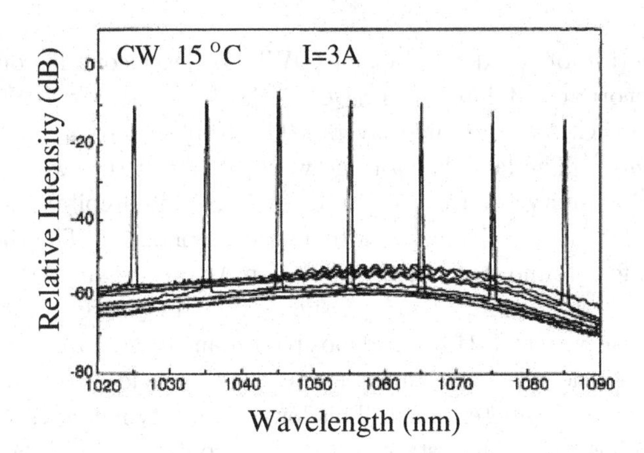

Fig. 6.38 Tuned emission spectra of external cavity laser between 1025 and 1085 nm at drive current of 3 A. Adapted with permission from *Electron. Lett.* **34**, 6, pp. 559. Morgott et al. (1998).

6.3.3 *Broadly tunable quantum dots lasers*

Quantum dots (QDs) have been called artificial atoms in its ultimate confinements in a box or a dot with the discrete energy levels of electrons. Semiconductor quantum well (QW) lasers in a grating-coupled external cavity have been widely used for their continuous tunability that is free of mode-hops and limited by the gain spectral width of the QW active medium. In contrast to QW lasers, QDs have the following advantages for tuning applications. (i) inhomogeneous broadening results in broad gain spectrum; (ii) low density of states leads to low threshold current J_{th} and carrier population of higher energy states at low pump current; (iii) homogeneous gain broadening enables sufficient gain to occur at selected frequency along with suppression of free-running lasing; (iv) broad tuning range is achieved at low bias.

An anti-reflection coated single-stack quantum dot laser in a grating-coupled external cavity has been shown to operate across a tuning range from 1.095 μm to 1.245 μm [Li *et. al.* (2000)]. This 150 nm range extends from the energy levels of the ground state to excited states. At any wavelength, the threshold current density is no greater than 1.1 kA/cm^2. Among other characteristics, a room temperature $J_{th} = 26\ A/cm^2$ and a 0.1 linewidth enhancement factor have been measured. These results suggest that numerous opportunities exist to use QD lasers as tunable coherent light sources.

The structure of the dots-in-a-well (DWELL) laser contains one InAs QD layer incorporated into an $In_{0.2}Ga_{0.8}As$ 10 nm thick QW and is sandwiched by GaAs waveguide layers. The in-plane density of QDs is $7.5 \times 10\ ^{10} cm^{-2}$. The laser has a ridge waveguide structure with a width of 9 μm and a cavity length of 2.0 mm. It is mounted epitaxial-side up on a heat sink that is stabilized at a temperature of 20 °C. The threshold current, J_{th}, for ground state lasing is I =35 mA without any AR-coating. From this data, $J_{th} = 194\ A/cm^2$. A single $\lambda/4$ HFO_2 layer designed for minimum reflectivity at 1.24 μm is deposited on one facet. From the difference in slope efficiencies between the two mirrors, a residual reflectivity of approximately 1.6% is determined. This low reflectivity increases the total cavity loss. Thus, the ground state lasing of the solitary laser is completely extinguished.

The emission spectra of the solitary device under different pump levels is shown in Fig. 6.39. With a bias as low as 200 mA (1.1 kA/cm^2), the

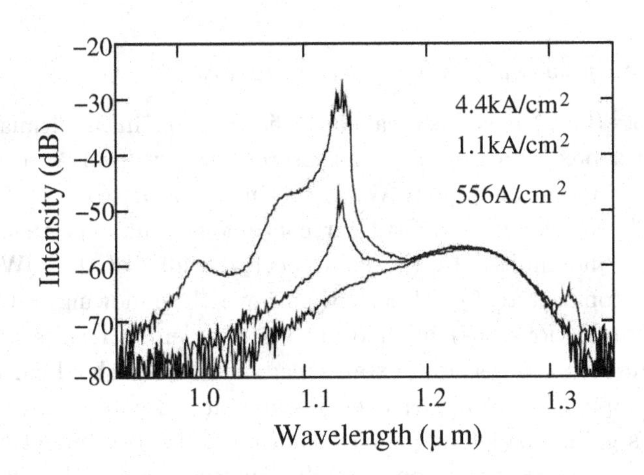

Fig. 6.39 electroluminescence spectrum of a DWELL laser for different pump levels. Adapted with permission from *IEEE Photo. Tech. Lett.* **12**, 7, 759-761. Li et al. (2000).

ground state at 1.24 μm is already well saturated and the first excited state is significantly populated. Since the AR-coating is not especially low, the device begins lasing under this pump level at the first excited state around 1.14 μm. The spectrum covered is from 1.1 $\sim$ 1.25 μm. Upon increase of the pump level to 800 mA (4.4 kA/cm^2), the second excited state starts filling at 1.07 μm, accompanied by obvious carrier filling of the lowest energy QW state at 1.0 μm. This spectrum spans a wavelength range of more than 200 nm.

A simple external cavity configuration that includes a collimator and a diffraction grating is subsequently constructed. Tuning was achieved by rotating the grating to select a certain wavelength emission to be reflected back to the laser. The laser is operated in pulsed mode with a pulse width of 500 ns and a duty cycle of 2%. The distance between the laser and the grating is about 25 cm, corresponding to a round trip time of $\sim$ 1.7 ns, which allows the photons to travel many round trips during the pump pulse time. In this case, pulsed mode operation is equivalent to CW pumping.

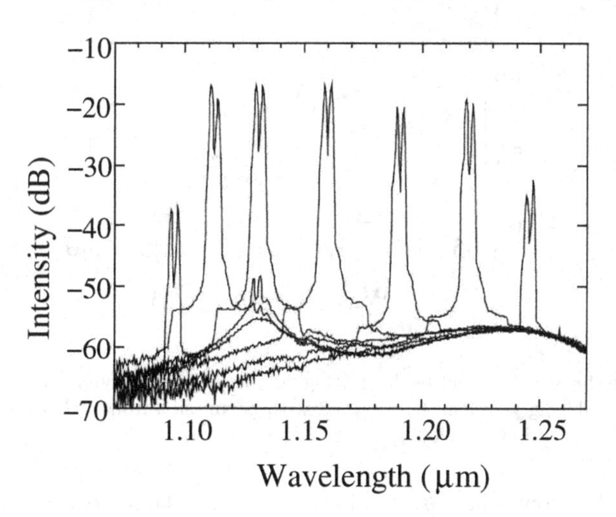

Fig. 6.40 Lasing spectra of grating-coupled external cavity DWELL lasers. When rotating the diffraction grating, the lasing wavelength is tuned across the wavelength range of 1.095 – 1.245μm. Adapted with permission from *IEEE Photo. Tech. Lett.* **12**, 7, 759-761. Li et al. (2000).

Figure 6.40 shows the actual tuning range achieved in the experiment by rotating the grating. At each step the spectrum is measured by an optical spectrum analyzer. Across most of 150 nm range, the lasing peak in the spectrum is higher than the pedestal spontaneous level by about 25$\sim$30 dB. The lasing linewidth is about 3 nm. A narrower linewidth could be obtained by using a Littman-Metcalf configuration for the external cavity and by fabricating a narrower ridge waveguide laser. Significantly, no failure of lasing occurs across the energy gap between the ground and the first excited state. This is, as expected, due to homogenous gain broadening. It is possible that the double peaks appearing in the lasing spectrum are due to the spatial modes of the laser. However, it cannot yet be ruled out that the twin peaks are evidence of inhomogeneous broadening, similar to what has been observed for the free-running broadband emission in quantum dot lasers at room temperature.

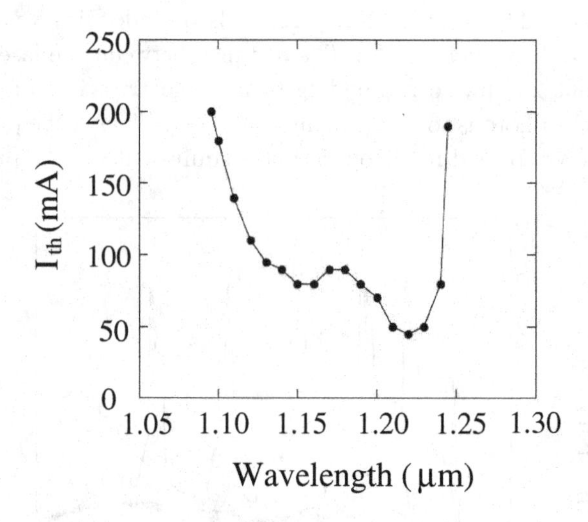

Fig. 6.41 The threshold current of the grating-coupled external cavity DWELL laser as a function of the lasing wavelength. Adapted with permission from *IEEE Photo. Tech. Lett.* **12**, 7, 759-761. Li et al. (2000).

Figure 6.41 shows the threshold current of the tunable external cavity DWELL laser as a function of the tuning wavelength. With no more than 200 mA bias (1.1 kA/cm^2), one can tune across 150 nm. At 1.18 μm, which is about halfway in the energy gap between the ground and first excited states, the threshold current increases only slightly. The lowest threshold current for the ground state is ~45 mA, which is higher than the original uncoated laser threshold. This increase is due to the fact that only about 10% of the power is reflected back into the device. Consequently the total loss is larger than that of the original uncoated laser. An increase of the external cavity feedback to a realistic 30% would decrease the threshold current and extend the tuning range further.

Figure 6.42 shows several typical L-I curves measured from the uncoated facet of the device at different wavelengths. The solid curve is for λ = 1.23 μm in the ground state. The dotted curve is measured for $\lambda = 1.17$ μm

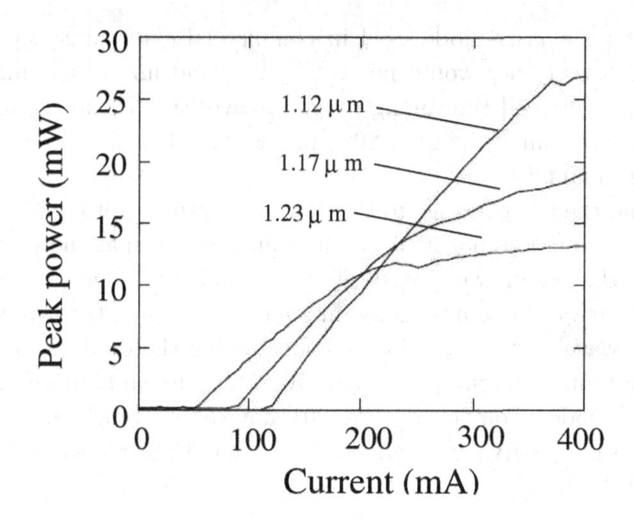

Fig. 6.42 Peak pulse power versus current for lasing at 1.23 μm (solid line), 1.17 μm (dotted line), 1.12 μm (dashed line). Adapted with permission from *IEEE Photo. Tech. Lett.* **12**, 7, 759-761. Li et al. (2000).

in the gap between the ground and the first excited states. The dashed curve is for $\lambda = 1.12\ \mu m$ at the short-wavelength shoulder of the first excited state. The peak power at $\sim 200\ mA$ is about $\sim 10\ mW$ for all three wavelengths, and the slope efficiency is approximately 0.1 W/A. Above 200 mA the slope efficiency decreases at $\lambda = 1.23\ \mu m$ and $\lambda = 1.17\ \mu m$ mainly because of the strong carrier filling at the first excited state under this pump level. A decrease in the reflectivity of the AR-coating will partially improve the slope efficiency and increase the linear range in the L-I curve. However, around the center wavelength of the first excited state, the slope efficiency remains constant for increasing pump level, as the dashed curve shows in Fig. 6.42 .

At 1.24 μm and 1.10 μm, a slope efficiency of 0.041 W/A and 0.066 W/A is measured, respectively, and a useful output power in the milliwatt range is still obtained. The tuning range and lasing behavior could be improved by decreasing the reflectivity of the AR-coating and modifying the external cavity design. From Fig. 6.39, the Fabry-Perot laser without external grating lases at a low pump current of ~200 mA. This device feature inhibits carrier filling of the higher energy states, therefore hindering the tuning range on the short wavelength side. If the residual reflectivity of the AR-coating at the first excited state was 5×10^{-4}, (which is a reasonable requirement for a laser diode used in commercial external-cavity lasers), the original laser cavity would not reach threshold until ~450 mA. This situation would extend the tuning range to cover the second excited state and make a total tuning range of 200 nm possible. This figure represents a 17% tuning capability.

To extend the tuning range to longer wavelengths is not trivial since increasing the pump does not increase the gain very much at the wavelength longer than the center wavelength of the ground state. Only a reduction in the total cavity loss or a longer gain region can extend the tuning range on the long wavelength side. However, decreasing the cavity loss will put a more stringent requirement on the AR-coating to prohibit the internal Fabry-Perot modes from lasing. A 201 nm tuning range in a grating-coupled external cavity quantum dots laser has been reported [Varangis *et. al.* (2000)].

Chapter 7

Frequency Stabilization of Tunable External Cavity Diode Lasers

7.1 Introduction

The development of stabilization of diode laser technology got started after that of gas and solid state lasers. At the earlier stage, frequency stabilization for diode lasers began with a scheme using Fabry-Perot interferometer as a frequency reference. The research on frequency stabilization for diode lasers made great progress in the 1980s when coherent optical telecommunications were proposed and demonstrated [Yamamoto (1980); Okoshi and Kikuchi (1981); Yamamoto and Kimura (1981)], and in the 1990s when when diode lasers found applications in basic research of atomic and molecular physics [Wieman and Hollberg (1991); Hemmerich *et. al.* (1990a); Maki *et. al.* (1993); Snadden *et. al.* (1997)]. Both require frequency stabilized diode lasers as well as precise frequency stabilization [Ikegami *et. al.* (1995)].

Stable and accurate frequency references can be used for single-point and linear scanning frequency calibration. There are many approaches to produce frequency references such as the use of elementary atomic, molecular, and solid state absorption lines, artifact references of interferometer etalon, and fiber gratings [Gilbert *et. al.* (2002)]. The fundamental references can provide very accurate calibration points. However, some convenient references are not available in all the wavelength regions, and artifact references suffer from strong dependence of stability on temperature, pressure, and stain. Therefore, some other stabilization techniques are developed by optical feedback and optical self-heterodyne beat-frequency control. We try to explore these techniques in this chapter.

7.2 Basic concepts of frequency stabilization

The output frequency of an external cavity diode laser depends on the injection current and the temperature as described in chapters 4 and 5. To obtain a stable frequency output, it is important to stabilize the diode's temperature and the injection current. In a diode laser with grating feedback the output beam reflects off the grating, while the first-order diffracted beam is directed back into the laser diode in Littrow configuration. The optical feedback from the grating is spectrally narrowed and peaked at a frequency that can differ from the central frequency of the free-running diode laser. The feedback narrows the laser linewidth to from 50 MHz to 1 MHz. The central frequency will be very close to that of the feedback signal. Many experiments require a laser with a well-defined frequency. But over time, the central frequency of a diode laser with grating feedback will drift, this drift is caused by fluctuations in temperature and injection current and mechanical fluctuations. Stabilizing the laser by locking it to an external reference reduces this drift. Laser frequency stabilization is based on the generation of a frequency error signal, which passes through zero at the locking frequency.

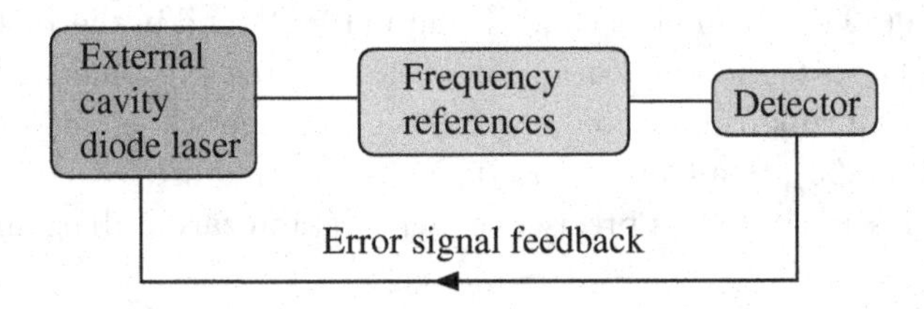

Fig. 7.1 Basic scheme for frequency stabilization of semiconductor diode laser.

To achieve the frequency stabilization of diode laser, one has to compare the operating frequency of diode laser with frequency reference. Fig. 7.1 illustrates the fundamental idea for stabilizing the frequency of semiconductor diode laser. A portion of output diode laser is compared with a frequency reference, then error signal is converted into the electrical signal and feedback to diode laser after amplification if necessary. There are a variety of reference frequencies which can be used to stabilize the frequency of diode laser. The most basic one is to use a Fabry-Perot interferometer as reference as schematically shown in Fig 7.2.

A laser is driven by DC current source, a sinusoidal AC current is also

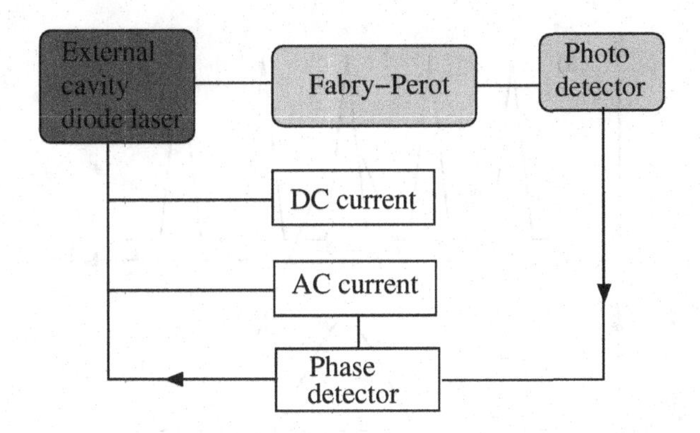

Fig. 7.2 Schematic diagram of frequency stabilization of semiconductor diode laser by use of Fabry-Perot interferometer.

applied to the laser from a local oscillator. The output beam from laser passes through the F-P interferometer, the transmission spectrum is shown in Fig 7.3(a). If the laser frequency is a little lower than the center frequency of one of the resonant peak of FP, as shown by frequency f_{-1} in Fig 7.3(b), the output signal is in phase with the initial modulation signal. If the laser frequency is a little higher than the center frequency of one of the resonant peak as shown by frequency f_1 in Fig 7.3(b), the output signal is out of phase with the initial modulation signal. If the laser frequency is equal to the center frequency of one of the resonant peak as shown by frequency f_0 in Fig 7.3(b), the output signal is cutoff by the low pass filter. The output signal from the photo detector is discriminated by the phase detector, and the error signal [Fig 7.3(c)] is feedback to the laser injection current. Therefore, the laser frequency is stabilized at the point of the center frequency of the resonant peak. In this way, the relative instability for the laser frequency was achieved $< 10^{-7}$ in a time interval of couple of minutes [Wieman and Gillbert (1982); Nakamura and Ohshima (1990)].

There are many candidates for frequency references for laser stabilization schemes, such as interferometers, etalons, fiber Bragg grating, atomic transitions, gas molecular absorptions, and ion absorption. Table 7.1 summarizes some of the frequency reference candidates.

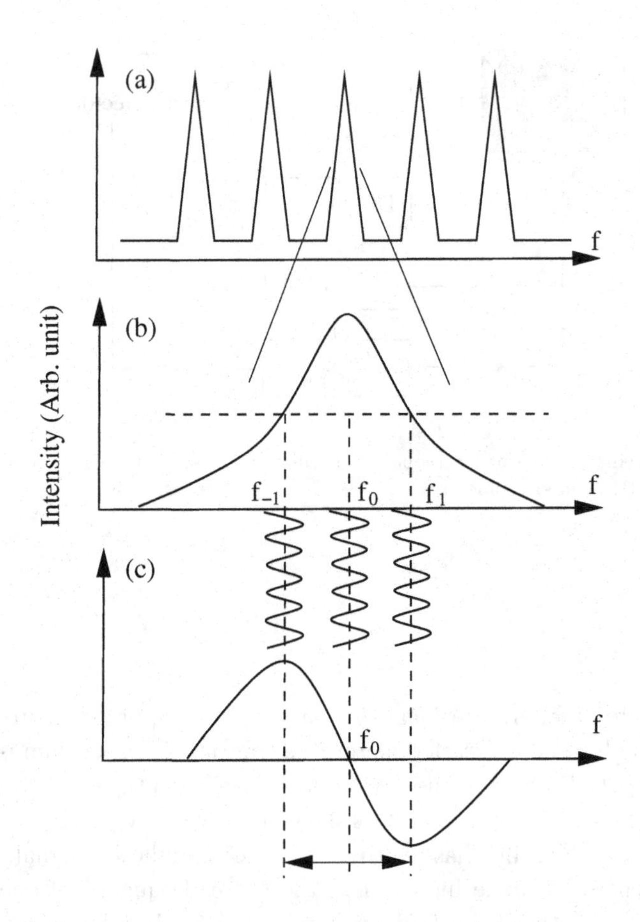

Fig. 7.3 (a)Transmission spectrum of F-P interferometer. (b) One of the F-P peak spectrum. (c) Generation of error signal.

7.3 Frequency stabilization schemes and apparatus

In this section, we introduce the frequency stabilization techniques by use of interferometers, atomic transition lines and gas molecular absorption, respectively.

7.3.1 *Interferometers: F-P etalon*

An interferometer, such as a Fabry-Perot Etalon, is the most commonly used frequency references for laser stabilization in scientific research and optical communication due to simple setup and wide tuning range. Fiber

Table 7.1 Summary of candidates for reference frequency.

Candidates	Type	Media	References
Interferometer & Etalon	Fabry-Perot, Waveguide resonator, FBG	Air, Quartz Glass	Nakamura et al. (1990); Bykovski (1970); Picque et al. (1975)
Atomic transitions	Normal absorption, Zeeman shift (DAVLL)	Rb, Cs	Corwin et al. (1998); Clifford et al. (2000); Morzinski et al. (2002)
Gas molecular absorption	Diatomic, Polyatomic molecular	$I_2, H^{13}C^{14}N, CO$	Swann et al. (2000); Conroy et al. (1999)
Ion absorption	Spectral hole burning	Er^{3+}:D^-:CaF_2, $Tm^{3+} : CaF_2$	Sellin et al. (1999); Bottger et al. (2003)

Bragg grating, planar waveguide filters, or any other kind of frequency selective filters can be used as frequency calibration if appropriate measures are taken. The first derivative of the Fabry-Perot transmission characteristics curve can be used as a good discrimination to lock the operation frequency of the laser diode as shown Fig. 7.3(c). If the laser frequency is modulated by adding a small sinusoidal component to laser injection current, and given by

$$\nu = \nu_0 + \nu_{FM} \sin(\omega t), \tag{7.1}$$

where ν_0 is the average laser frequency, ν_{FM} is the frequency modulation amplitude, ω is the modulation frequency.

The light transmission through an ideal etalon can be expressed by

$$T = \frac{T_{max}}{1 + F \sin^2 \phi}, \tag{7.2}$$

where $\phi = 2\pi(\nu - \nu_m)nd \cos\theta/c$, the maximum transmission occurs at $\phi = q\pi$, where q is positive integer, T_{max} is the peak etalon transmission, F is the etalon finesse, nd is optical path length of etalon, ν and ν_m denote the laser and the nearest etalon transmission peak frequency, respectively. If $\phi \ll 1$, that is

$$\phi = 2\pi(\nu - \nu_m)nd \cos\theta/c \ll 1, \tag{7.3}$$

then the transmission can be expanded in a Taylor series around the average laser frequency ν_0, the first-order approximation of transmission T is

$$T(\nu) = T(\nu_0) + T'(\nu_0)\nu_{FM} \sin(\omega t), \tag{7.4}$$

under the assumption of $(\nu - \nu_m) \ll c/(2ndF\cos\theta)$. Then the etalon transmission function and its first derivative are written as to a good approximation

$$T = T_{max}, \tag{7.5}$$

and

$$T'(\nu_0) = T_{max}[-8(\nu_0 - \nu_m)(\frac{2ndF\cos\theta}{c})^2]. \tag{7.6}$$

Eq. (7.6) presents the first derivative of etalon transmission curve with respect to laser frequency. It goes to zero when $\nu_0 = \nu_m$, i.e., the laser frequency is locked, and change sign whenever $\nu_0 - \nu_m$ changes sign as shown by Eq. (7.4). Therefore, this is a convenient error signal for the feedback loop of frequency stabilization.

The first experiment of frequency stabilization using Fabry-Perot etalon was done by Bykovski et al. [Bykovski (1970)]. Picque and Roizen [Picque and Roison (1975)] made use of the dependence of the output frequency on the diode injection current to lock a cw single-mode tunable GaAs diode laser to an external passive cavity. The resulting long-term stability of the laser optical frequency was checked by making a reference to a microwave frequency standard in terms of the light-shift effect in a cesium vapor. The laser frequency was found to track the drift of the cavity bandpass with residual variations less than 300 kHz rms, servo-controlled frequency scanning was achieved over 5 GHz.

7.3.2 *Atomic transition line*

Atomic transitions can offer good frequency references in the wide wavelength range and stable frequency references in various environmental conditions (temperature, strain, and pressure). Transitions of hyperfine splitting of energy levels of the alkali like rubidium and cesium are commonly used as frequency references [Arditi and Picque (1975); Hori *et. al.* (1983); Sato *et. al.* (1988)].

7.3.2.1 *Saturated absorption*

One of the popular method of stabilizing the diode laser is saturated absorption spectroscopy, the schematic experimental setup is shown in Fig. 7.4. A narrow peak in a saturated absorption spectrum is often used as an external reference for frequency locking a diode laser [Kruger (1998)] , a fraction

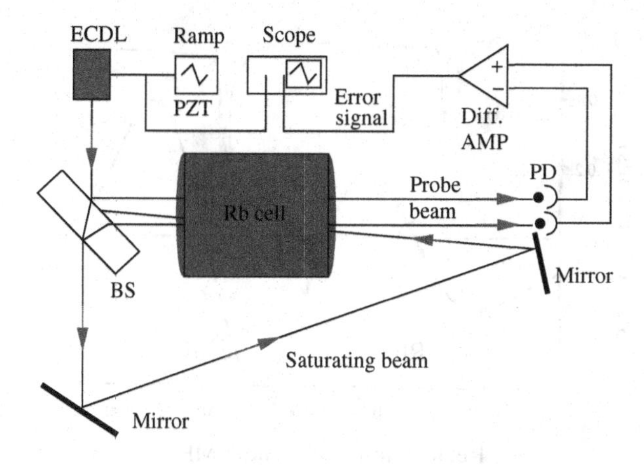

Fig. 7.4 Saturated absorption experimental setup. ECDL: external cavity diode laser, BS: beam splitter, PD: photodetector.

of the laser output is sent through an atomic vapor cell. A saturated absorption spectrum is measured and the laser frequency is tuned to either the side or the peak of a narrow, saturated absorption line as illustrated in Fig. 7.5(a). Side-locking is one of the simplest stabilization methods. On the side of a narrow absorption line the output voltage V(ν) of the differential photo-detector has a steep slope as a function of the laser frequency ν. To lock the laser to a frequency ν_o, for which $dV(\nu)/d\nu|_{\nu_o} \neq 0$, a reference voltage V($\nu_o$) is subtracted from the output signal to produce an error signal err(ν) $= V(\nu) - V(\nu_o)$. This error signal serves as an input to a feedback loop which adjusts the laser's frequency to produce error(ν) $= 0$. This is accomplished by adjusting the PZT voltage. Side-locking is widely used in laser-cooling of neutral atoms where a small detuning of the laser frequency is necessary. A disadvantage of side-locking is that fluctuations in beam alignment and intensity will alter the lock point and cause a drift in the laser frequency.

Peak-locking is less sensitive to these fluctuations. But on a peak $dV(\nu)/d\nu|_{\nu_o} = 0$. If the laser frequency has been adjusted to equal the peak frequency ν_o, and the voltage of the differential photo-detector is being monitored, it is easy to detect a drift in the laser frequency. A drift towards higher or lower frequency causes a decrease in the output voltage and an error signal of the same sign. A non-zero error signal alone is therefore insufficient to determine whether the laser frequency should be increased or decreased.

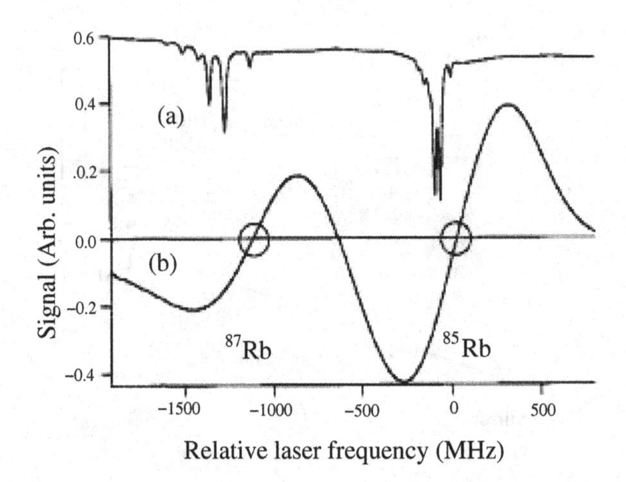

Fig. 7.5 (a) Saturated absorption spectrum of Rb. (b) DAVLL signal as the diode laser is scanned across Rb resonances with PZT. A laser can be locked to either of the two circled zero crossings of the DAVLL signal. These feature are arising from the transitions of ^{87}Rb F=2 $\rightarrow$ F$'$=1,2,3 and the ^{85}Rb F=3 $\rightarrow$ F$'$=2,3,4. the frequency of locking point can be tuned optically by rotating the quarter wave plate, or electronically by adding an offset voltage to the signal. Adapted with permission from *Appl. Opt.* **37**, 15, pp. 3295-3298. Corwin et al. (1998).

To lock onto a peak, the laser frequency is dithered slowly at a frequency Ω. An AC signal with frequency Ω in the kHz range is fed into the reference channel of a lock-in amplifier and into the controller of the PZT. The PZT expands and contracts, which changes the length of the laser cavity. The frequency of the laser light as a function of time varies as

$$\nu(t) = \nu_o^{'} + \Delta\nu \cos(\Omega t). \tag{7.7}$$

Here $\nu_o^{'}$ is the frequency of the laser when it is not dithered and $\Delta\nu \gg \Omega$; $\nu_o^{'}$ may differ from ν_o if the laser frequency has drifted. The output voltage of the differential photo-detector thus varies as

$$V(t) = V(\nu_o^{'}) + \Delta\nu \cos(\Omega t). \tag{7.8}$$

The output voltage of the differential photo-detector becomes the input signal of the lock-in amplifier. The lock-in amplifier multiplies the reference signal with the detector signal and outputs a DC signal, which is proportional to the time average of product. A disadvantage of the dither-locking method is that the output of the laser is modulated directly, or that expensive electro-optic components must be used to modulate only the light entering the saturated absorption cell. Furthermore, both peak and side-locking techniques have a limited recovery range, since the resonance peaks are very narrow.

A technique to stabilize in an atomic transition the chirping frequency of a narrow semiconductor diode laser has been presented [Morzinski *et. al.* (2002)]. The technique has been demonstrated to chirp-cool ^{85}Rb atoms used for loading a magneto-optical trap. The stabilization process eliminates the long-term fluctuations and drifts in the number of atoms caught in the trap. This is a simple, easy-to-implement, and robust method for wide range of laser cooling experiments employing frequency chirping. Moreover, this technique allows stabilization the chirp for several continuous hours.

7.3.2.2 *Dichroic-Atomic-Vapor Laser Lock (DAVLL)*

The Dichroic-Atomic-Vapor Laser Lock (DAVLL) technique aims to extend the recovery range by employing a weak magnetic field to split the Zeeman components of an atomic Doppler-broadened absorption signal and then generating an error signal that depends on the difference in absorption rates of the two components. This technique was first demonstrated with a laser in helium [Cheron *et. al.* (1994)] and then developed at 780 nm for rubidium [Corwin *et. al.* (1998)], recently 895 nm for cesium [Clifford *et. al.* (2000)]. The Zeeman effect removes the degeneracy of atomic hyperfine states. A magnetic sublevel characterized by the quantum number m_F is shifted in energy by $\Delta E = g_F \mu_B B_0 m_F$, where B_0 is the magnitude of the external magnetic field, g_F is the Lande g-factor, μ_B is the Bohr magnetron. The shift in the transition energy for two sublevels is then given by

$$\Delta E_{trans} = \Delta E^{'} - \Delta E = \mu_B B_0 (g_{F'} m_{F'} - g_F m_F), \tag{7.9}$$

where the primed symbols are related to the upper state. To use the DAVLL technique, a small fraction of the laser light passes through an atomic vapor cell, the cell is placed inside a large solenoid. The magnetic field B generated by the solenoid is parallel or anti-parallel to the wave vector k of the laser light, the laser light must be linearly polarized.

Let B_0 point into the z-direction in which the light is propagating. The Zeeman effect splits each formerly-degenerate hyperfine energy level characterized by the quantum number F into 2F + 1 components characterized by m_F, with m_F ranging from F to -F in integer steps. For optical transitions the selection rules are Δm_F =0, 1. For $\Delta m_F = 0$ the electric field vector must be parallel to the magnetic field B_0. But the electric field vector E of the laser light oscillates in a plane perpendicular to B_0, so no $\Delta m_F = 0$ transitions are induced. Right-circular polarized light travelling anti-parallel to B_0 induces $\Delta m_F = 1$ transitions, and left-circular polarized light travelling anti-parallel to B_0 induces $\Delta m_F = -1$ transitions.

The Zeeman effect shifts the transition energies for $\Delta m_F = 1$ transitions relative to the transition energies for $\Delta m_F = -1$ transitions. If the Doppler broadened absorption curve is shifted towards higher frequencies for right-circular polarized light, then it is shifted towards lower frequencies for left-circular polarized light.

After the light has passed through the vapor cell, it is split into two beams with orthogonal circular polarizations. This can be done using a quarter-wave plate and a linear polarizing beam splitter. The quarter wave plate changes the two orthogonal circular polarization components in the beam leaving the vapor cell into two orthogonal linear polarization components. The linear polarizing beam splitter directs these component into different photodiode detectors. The output voltages of the two detectors are proportional the intensities of the right and left hand circular polarized beams exiting the cell. By subtracting the two output voltages an anti-symmetric error signal is generated which passes through zero and is suitable for locking. The advantage of the DAVLL technique over the side-locking or peak-locking techniques is its large tuning range. The tuning range of the DAVLL technique is limited by the width of the Doppler broadened absorption peaks, while the tuning range of the other techniques is limited by the width of the Doppler-free saturated absorption peaks. Frequency modulation is not required.

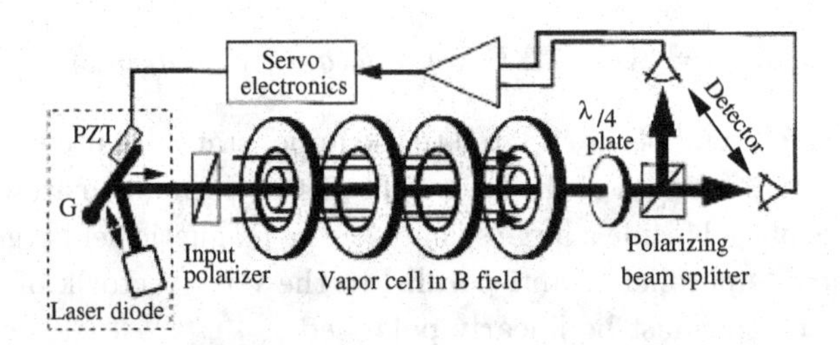

Fig. 7.6 Schematic of a DAVLL setup. G: grating. Adapted with permission from *Appl. Opt.* **37**, 15, pp. 3295-3298. Corwin et al. (1998).

Corwin et al. [Corwin *et. al.* (1998)] have demonstrated the robust aforementioned method to stabilize a diode laser frequency to an atomic transition. Experimental setup is schematically shown in Fig. 7.6, diode laser system is in typical Littrow configuration. The SDL 780-nm diode laser is tuned by use of a diffraction grating, the output beam from this laser passes through a beam splitter, and a small amount of power is split off to be used for locking. After passing through a small aperture, the resulting beam passes through a linear polarizer. Pure linear polarization is equivalent to a linear combination of equal amounts of two polarizations. This beam next passes through a cell-magnet combination, consisting of a glass cell filled with Rb vapor and a 100-G magnetic field. To generate the DAVLL signal, the absorption profiles of the σ_+ light must be subtracted from that of the σ_-. To this end, the two circular polarizations are converted into two orthogonal linear polarizations by passing through a quarter wave plate. The two linear polarizations are separated by a polarizing beam splitter, and the resulting two beams are incident on two photo-detectors whose photocurrents are subtracted. As the frequency of laser is scanned across an atomic transition, asymmetrical curve is generated, as shown in Fig. 7.5(b) and Fig. 7.7 . The diode laser is then locked by feeding back a voltage to the PZT so that a DAVLL signal is maintained at the central zero crossing. The frequency of a 780 nm diode laser has been stabilized within the drift of less than 0.5 MHz peak to peak (1 part in 10^9) over a period of 38 hours.

The use of DAVLL in Cs vapor to stabilize an ultra-compact extended cavity diode laser at 852 nm has been demonstrated [Clifford *et. al.* (2000)]. The expected laser stabilization error signal for a range of magnetic fields has been investigated and the ability to tune the locked ECDL by variation of the magnetic field has been achieved. The ECDL has a linewidth of 520 kHz and the drift when locked is of the order of 5 MHz/h, which is more than an order of magnitude improvement on unlocked laser system.

Recently, a new design of diode laser frequency stabilization system using the Zeeman effect has been presented [Yashchuk *et. al.* (2000)]. A wide range of regimes of operation are analyzed, this system is different from the original design of JILA [Corwin *et. al.* (1998)] in that the magnetic fields from cell-magnet system are fully contained and thus it can be used in proximity of magnetically sensitive instruments. It is shown that the simple readjustment of the respective angles of optical elements allows one to extend the frequency tuning range to the wings of a resonance line. Baluschev [Baluschev *et. al.* (2000)] demonstrated a new scheme of frequency locking of a diode laser to a two-photon transition of rubidium atomic line by using the Zeeman modulation technique. Frequency of the diode laser has been

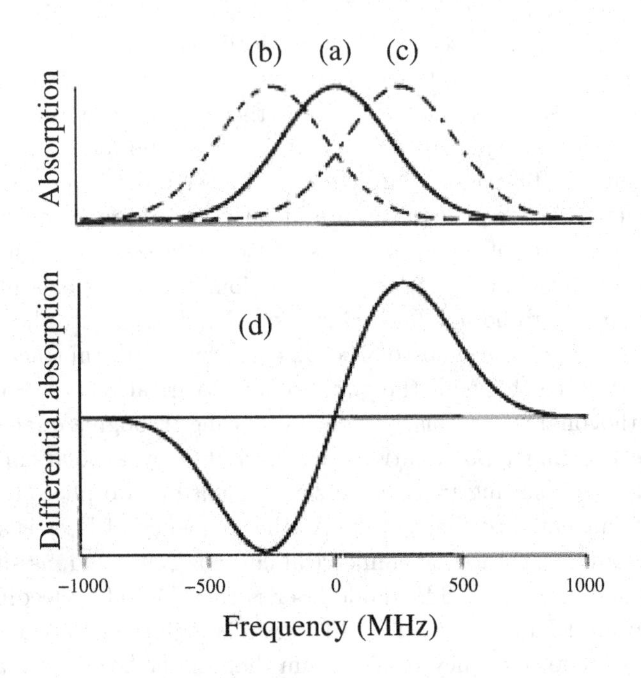

Fig. 7.7 Original of the DAVLL error signal. (a) A Doppler broadened transition in Rb in the absence of magnetic field. (b) The same transition, Zeeman shifted in a 100 G magnetic field, when circularly polarized light is incident on the vapor. (c) The same as (b), but with opposite circular polarization. (d) The difference between (c) and (b) giving the DAVLL signal. Adapted with permission from *Appl. Opt.* **37**, 15, pp. 3295-3298. Corwin et al. (1998).

locked and tuned by modulating and shifting the two-photon transition frequency with ac and dc magnetic fields. A narrow linewidth of 500 kHz and continuous tunability over 280 MHz without laser frequency modulation.

7.3.3 *Gas molecular absorption*

Many candidates of gas molecular absorption line spectra can be used as a frequency discriminator for the laser stabilization [Swann and Gilbert (2000)]. The frequency of a 1.5 μm external-cavity semiconductor laser has been stabilized by detecting an acetylene (C_2H_2) absorption line with an offset-tracking DBR laser. Absolute stability better than 200 kHz is achieved while a linewidth narrower than 140 kHz and a continuous tuning range of 4 GHz are retained [Ishida and Toba (1991)]. In contrast with other candidates, gas molecular as frequency reference have more favorable advantages as follows: (i) molecular vibration-rotation transition have narrow natural linewidth; (ii) absorption can be obtained thermally populated ground states so that perturbing effects can be eliminated; (iii) the high density of molecular spectra results in high probability of molecular excitation by a laser field; (iv) molecules with symmetry and insensitive to Stark and Zeeman shifts are suitable for candidates; (v) the pressure shift is about two orders of magnitude smaller than that of atomic transition lines.

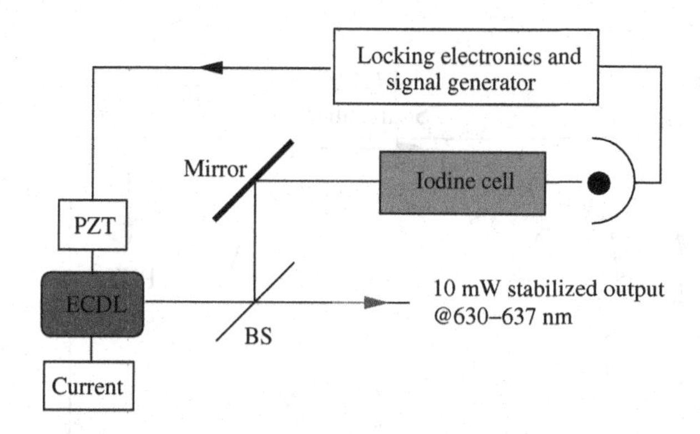

Fig. 7.8 Schematic of the stabilized laser configuration. BS: beam splitter, ECDL: external cavity diode laser, PZT: piezoelectrical transducer. Adapted with permission from *J. Mod. Opt.* **46**, 12, pp. 1787-1791. Conroy et al. (1999).

A compact, high-performance extended-cavity diode laser in the Littrow geometry has been realized at 635 nm with an upper limit of 470 kHz on the laser linewidth and obtain an output power of more than 10 mW. The laser may be tuned continuously over 21.6 GHz and discontinuously from 630 to 637 nm. Spectroscopy of iodine[Arie *et. al.* (1992)] has been performed and the laser stabilized to a molecular absorption feature [Conroy *et. al.* (1999)]. The laser diode used for this system is a SDL 7501-G1, giving a nominal output power of 15 mW at 635 nm. The diode is anti-reflection coated on its front facet ($R < 0.01\%$). The laser is placed in an extended cavity Littrow geometry with a 15 mm^2 1200 lines/mm diffraction grating, blazed for 300 nm. The grating is placed 10$\pm$20 mm from the diode facet and reflected 20% of the incident light back into the laser. The zeroth order output of the grating gives the output used for the experiment.

Fig. 7.8 shows a schematic of the laser system used. The diffraction grating is mounted in a machined holder inserted into a clear quadrant design commercial mirror mount. The laser diode is mounted in a commercial collimating tube. These components are mounted on a baseplate, which was temperature stabilized using a thermoelectric cooler. The temperature itself could be varied over a 30 °C range. The system is isolated from mechanical vibrations using Sorbathane and enclosed. The diode is operated in constant current mode with an input current of up to 70 mA with a maximum output power of 10.85 mW. One records an instrument-limited ECDL linewidth of 470 kHz for the system in a 10 ms timescale.

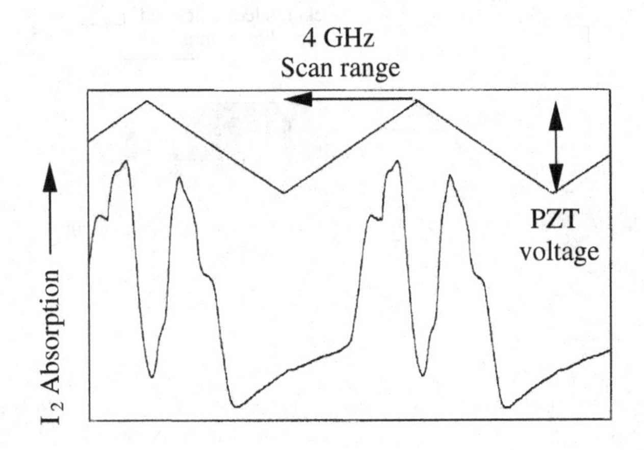

Fig. 7.9 Trace of I_2 absorption spectrum around 15786 cm^{-1}. Adapted with permission from *J. Mod. Opt.* **46**, 12, pp. 1787-1791. Conroy et al. (1999).

This laser system is used to perform spectroscopy of iodine for the purpose of finding a frequency reference. The absorption lines of the I_2 $B-X$ system are commonly used for the frequency stabilization of many laser systems, most notably HeNe lasers. The doppler-broadened absorption lines are broad, typically more than 1 GHz (FWHM) and of low intensity, limiting their ability to give ultra-high frequency stability, however their frequency makes them ideal for discontinuously tunable sources. To lock the laser system to an iodine feature a fraction of the output (5%) is sent into an iodine cell, as a probe beam. The cell is 10 cm long and a single photodiode monitored the transmission of this probe.

Fig. 7.9 shows iodine absorption spectra of the lines at 15786 cm^{-1}. The two peaks can be clearly resolved with an indication of the underlying hyperfine structure. A simple side-of-fringe locking circuit was used to stabilize the laser system to the absorption line. The locking circuit bandwidth is 100 Hz and is therefore prone to high frequency noise that could cause the laser to jump out of lock. However, in quiet laboratory conditions, this laser was locked for in excess of one hour, and the long term drift of the lasing wavelength is eliminated.

7.3.4 *Persistent spectral hole burning*

The first demonstration of laser stabilization directly to a solid-state persistent hole burning material as a frequency reference has been reported by Cone group [Sellin *et. al.* (1999)]. Unlike the gas phase transition frequency reference and interferometer etalon frequency references as discussed in the previous sections, spectral hole burning can be prepared at any frequency within a broad inhomogeneous absorption profile. Such a frequency reference could be used as a programmable and transportable secondary frequency standard. Many scientific and device application requires highly stabilized laser source, lasers stabilized to spectral holes have found wide applications in ultrahigh resolution spectroscopy, optical processing, and wavelength-division multiplexing, etc. In addition, several lasers can be stabilized to multiple spectral holes, either in the same or separate absorption bands, with arbitrary frequency separations.

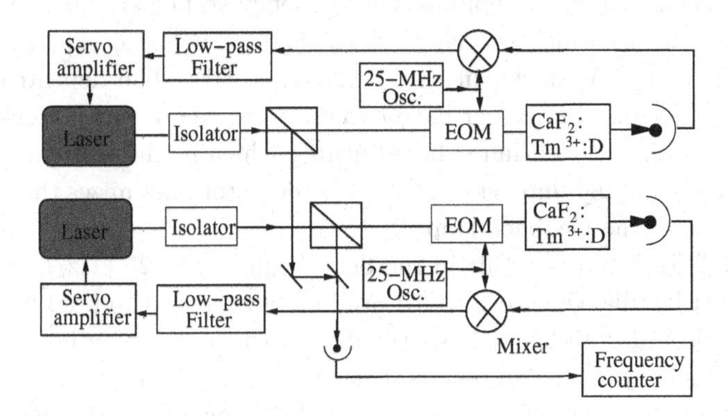

Fig. 7.10 Experimental setup for frequency locking to spectral holes and beat-frequency characterization of relative laser stability. EOM'S: electro-optic phase modulators. Adapted with permission from *Opt. Lett.* **24**, 15, pp. 1038-1040. Sellin et al. (1999).

The spectral hole-burning material for first demonstration was deuterated CaF_2:Tm^{+3}, other candidates have also been used [Bottger *et. al.*

(2003)]. The nominal Tm^{+3} concentration is 0.05% and linear absorption is 60% with persistent spectral hole burning width of 25 MHz on the $^3H_6 \rightarrow ^3H_4$ transition at wavelength of 798 nm. The holes have been shown to be persistent without measurable degradation for at least 48 h at 1.7 K. The experimental setup is shown in Fig. 7.10, two GaAlAs external cavity diode laser in Littman-Metcalf configuration are externally modulated with electro-optic modulators at 23 and 25 MHz for frequency locking to the spectral holes with Pound-Drever-Hall technique [Pound (1946); Drever *et. al.* (1983)]. This technique used frequency modulation spectroscopy to provide a corrective feedback signal to a servo amplifier that rapidly modulate the laser current and slowly adjust a peizoelectrically driven grating for optical feedback to keep current servo within operating limits. Each laser is independently locked to a spectral hole in a separate crystal. A single cryostat holds both crystals immersed in s superfluid helium at 1.9 K for improved thermal equilibrium and temperature stability. The stability of the frequency difference between the two lasers is measured by heterodyne detection and monitored on a frequency counter. The frequency of the heterodyne beat signal could be chosen arbitrary by the choice of the relative frequencies of the two spectral holes.

The spectral holes are shown in Fig. 7.11(a) and (b), since the spectra holes are broader than the free running laser jitter on this time scale, this procedure is carried out without active frequency stabilization and formed the initial holes to which the lasers are locked. The error signal for the hole in Fig 7.11(a) is shown in Fig. 7.11(c) and (d). The dependence of locking stability on laser power have two opposite factors: low intensity in the crystal minimizes continued hole burning which modifies the reference spectral hole; and high intensity incident on detector maximizes the signal-to-noise ratio of the feedback loop.

Fig. 7.12(a) displays a drift in the beat frequency of 25 kHz/min when the full laser irradiance used to burn original hole is used to keep the lock. For comparison, Fig. 7.12(c) shows the drift with free-running laser. Additionally, frequency drifts can be arised from unintentionally locking slightly off the center of hole. To minimize the drift one can reduce the laser intensity and therefore slow the rate of continuous burning as shown in Fig. 7.12(b). The degree to which the laser can be stabilized depends on the signal-to-noise ratio of the detection, feedback systems, and on the slope of the error signal determined by the depth and width of the hole. Therefore a narrow resonance hole burning width provides stronger stability of

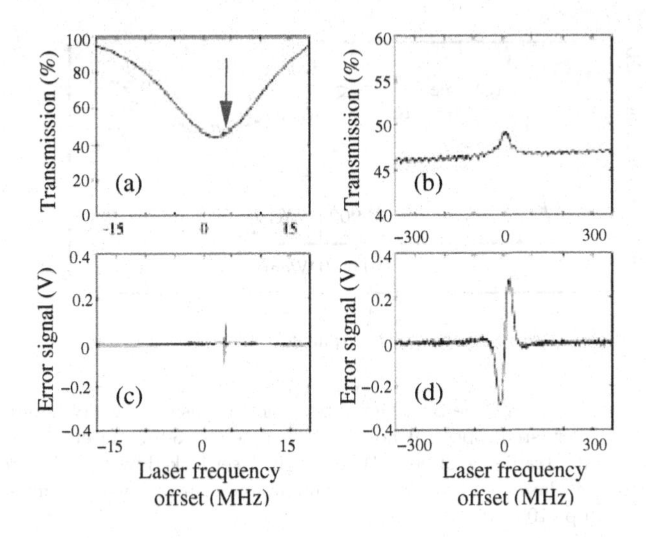

Fig. 7.11 Error signals derived from a shallow persistent spectral hole burned on the $^3H_6 \rightarrow ^3 I_4$ transitions of deuterated $Tm^{3+} : CaF_2$: (a) Transmission over the entire inhomogeneously broadened absorption profile and (b) transition over a smaller frequency range; (c),(d) error signals over the corresponding frequency ranges. The arrow in (a) indicates the spectral hole. Adapted with permission from *Opt. Lett.* **24**, 15, pp. 1038-1040. Sellin et al. (1999).

laser frequency.

More recently, the persistent spectral hole burning has been used as programmable laser frequency references to the important 1.5 μm optical communication window for the first time [Bottger *et. al.* (2003)], diode laser frequency stability of 2 kHz to 680 kHz over 20 ms 500 s has been demonstrated in the inhomogeneously broadened $^4I_{15/2} \rightarrow ^4 I_{13/2}$ optical absorption of $Er^{3+} : D^- : CaF_2$ with inexpensive diode laser. The system is isolated from vibrational or acoustical disturbance to spectral hole frequency reference, thus provides versatile, compact, stable source for many applications, for example, optical communications and computing, in either terrestrial or satellite systems.

7.4 Other frequency stabilization schemes

We have examined the traditional methods for frequency stabilization by using feedback to minimize laser frequency offset from an absolute frequency reference, such as atomic and molecular absorption lines, or a reference

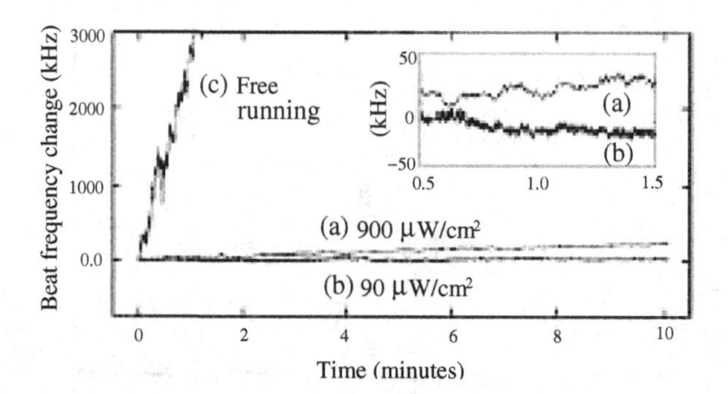

Fig. 7.12 Change in heterodyne beat signal between two lasers that are independently locked to separate persistent spectral holes and sampled at 50 ms intervals. (a) Laser locked with moderate irradiance of 900 $\mu W/cm^2$; (b) laser locked with tenfold reduction in irradiance; (c) both lasers free running for comparison. Adapted with permission from *Opt. Lett.* **24**, 15, pp. 1038-1040. Sellin et al. (1999).

interferometer. The compatibility of these stabilization approaches with frequency agility is determined by the frequency agility of the absolute reference. For application in which tuning is important, no-reference or self-referenced linewidth stabilization are especially attractive. Self-reference could be accomplished by comparing the past laser output as a reference with present output, then determining whether the present frequency has drifted or not and therefore deciding correction. Comparison between present and past laser output can be done by large path-difference interferometers. Several studies have used this technique to provide self-frequency stabilization [Chen (1989)]. Greiner et al. [Greiner *et. al.* (1998)] proposed and demonstrated a novel frequency-stabilization scheme. The method relies on sensing and control of a heterodyne beat signal derived from a fiber interferometer and functions in the absence of a fixed reference frequency. The stabilization method is shown to suppress impressed laser frequency modulation by nearly 2 orders of magnitude. Their method is well suitable for the systems in which both frequency agility and narrow linewidth are important.

Fig. 7.13 shows schematically the stabilization scheme. The output of a tunable diode laser is split and coupled into a Mach-Zehnder interferometer with one short and one long arm. Light passing through the short arm is acoustically frequency shifted by $\Delta\nu_{AOM}$. At the output port of the interferometer the light fields from short and long arms , E_s and E_l,

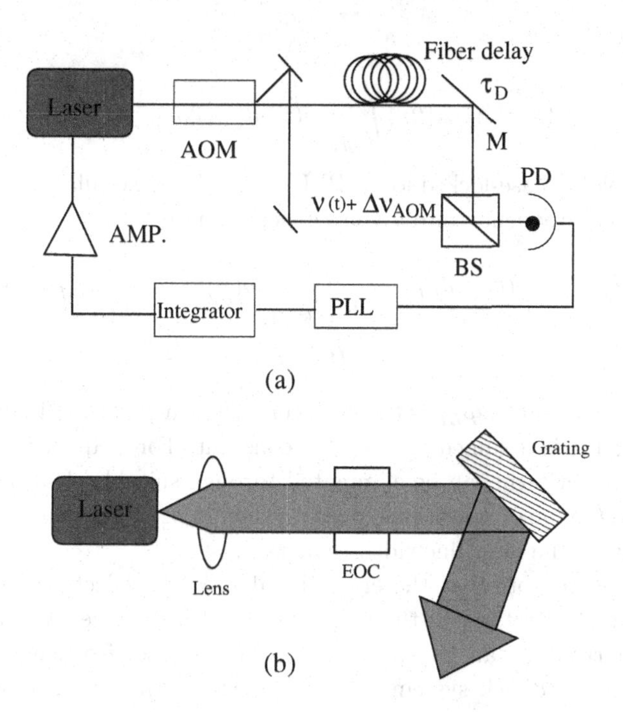

Fig. 7.13 (a) Schematic diagram of the apparatus; AMP: amplifier; PD: photo detector, BS: beam splitter. (b) Schematic of the short-external-cavity diode laser. EOC: electro-optic crystal. Adapted with permission from *Opt. Lett.* **23**, 16, pp. 1280-1282. Greiner et al. (1998).

respectively, can be written as

$$E_s = \epsilon(t)\exp[-j2\pi\int_{t_r}^{t}\nu(t')dt' - j2\pi\Delta\nu_{AOM}(t-t_r) - j\eta], \tag{7.10}$$

and

$$E_l = \alpha\epsilon(t-\tau_D)\exp[-j2\pi\int_{t_r}^{t-\tau_D}\nu(t')dt'], \tag{7.11}$$

where $\epsilon(t)$ is the real-valued amplitude of E_s, $\alpha(t)$ is the real-valued constant, τ_D is the fiber delay time, t_r is a reference time, $\nu(t)$ is the instantaneous frequency of the electric field before the AOM, and η represents the phase of the AOM at t_r. Addition of the two fields at the output of the interferometer produces an interferometer term in the detected signal given by

$$I_{beat} \propto 2\alpha\epsilon(t-\tau_D)\cos\{2\pi[\int_{t-\tau_D}^{t_r} \nu(t')dt' + \Delta\nu_{AOM}t + \eta]\}. \qquad (7.12)$$

One can exploit phase-locked loop (PLL) as a FM demodulator to convert the beat-signal frequency into a voltage according to

$$V_{PLL}(t) \propto \nu_{beat}(t) - \nu_{PLL} = \frac{d}{dt}\int_{t-\tau_D}^{t} \nu(t')dt' + \Delta\nu_{AOM} - \nu_{PLL}$$

$$= \nu(t) - \nu(t-\tau_D) + C., \qquad (7.13)$$

where C. is a constant, ν_{PLL} is the quiescent frequency of the PLL reference oscillator and it is assumed that $\epsilon(t)$ = constant. For frequency stabilization the constant term can be eliminated through suitable electronic filter. Under a wide range of conditions, one can then inject V_{PLL} into a feedback loop to provide the laser linewidth reduction.

It is worth to note that the error signal $V_{PLL}(t)$ reflects the laser frequency's time history rather than the absolute laser frequency. Therefore, linewidth narrowing can be applied at arbitrary laser frequency without adjustment of feedback system as is necessary in the interferometer frequency references. When $\nu(t)$ changes slowly over a fiber delay time τ_D, the error signal is proportional to the derivative of the laser frequency. In specific limits, subsequent integration electronics will yield a voltage that is proportional to the overall laser frequency excursion during the integration time, which may be useful for absolute frequency control. Control of $\Delta\nu_{AOM}$ provides a useful means of laser frequency adjustment. In the presence of large feedback gain, the error signal $V_{PLL}(t)$ relations zeroed and the following condition is met

$$\nu(t) - \nu(t-\tau_D) \approx \nu_{PLL} - \Delta\nu_{AOM}. \qquad (7.14)$$

Thus each value of $\Delta\nu_{AOM}$ corresponds to a specific time rate of change of laser frequency. When $\Delta\nu_{AOM} = \nu_{PLL}$, the laser frequency will keep stable.

The experimental setup is shown in Fig. 7.13. Laser is in typical Littrow configuration, the external cavity contains an antireflection-coated lithium tantalate electro-optic crystal configured as a transverse phase modulator that provides rapid frequency tuning. The long arm of fiber interferometer consists of a 2.26-km-long fiber segment yielding a time delay $\tau_D \approx 11$ μs. ν_{PLL} is set to be approximately 75 MHz so that it matches the frequency of $\Delta\nu_{AOM} \approx 75$ MHz. The output of PLL is subsequently connected to a

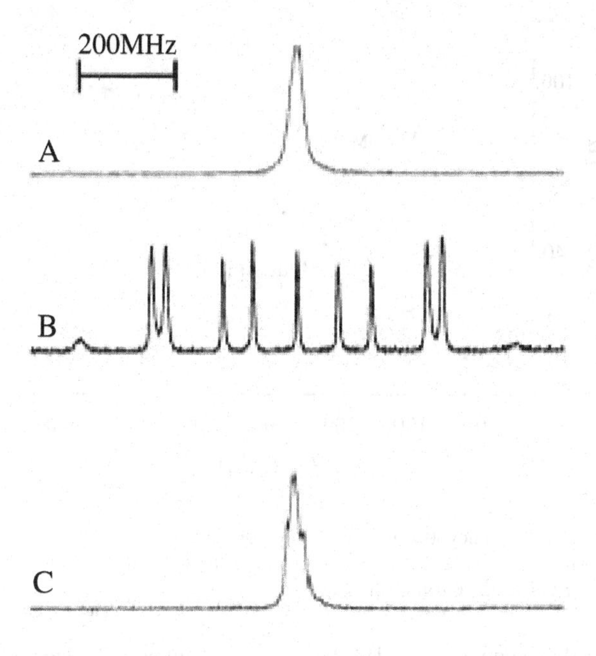

Fig. 7.14 (a) Trace A, optical power versus frequency for the quiescent laser. Trace B, broadened power spectrum after introduction of FM noise by laser diode current modulation. Trace C, power spectrum of the stabilized laser. Adapted with permission from *Opt. Lett.* **23**, 16, pp. 1280-1282. Greiner et al. (1998).

passive bandpass filter via an op-amp, then as a negative feedback to the electro-optic crystal. Fig. 7.14 shows single-event optical power spectra of the laser measured with 50-cm confocal cavity. Trace A is the spectrum of the free-running laser with a mean FWHM linewidth of 3.7 MHz. Trace B displays the spectrum of laser when diode current is modulated at 1 kHz. Next, the heterodyne beat feedback circuit is closed, and the laser spectrum of trace C is attained. The stabilized linewidth has a mean of 4.8 MHz (FWHM). Fig. 7.15 demonstrates the laser tuning by control of $\Delta\nu_{AOM}$ and shows how the beat signal's frequency changes at a rate of 0.2 MHz/μs for a duration of 500 μs when $\Delta\nu_{AOM}$ is changed by 3 MHz.

A simple method for long term stabilization and tuning of a diode laser which is locked to an external cavity by optical feedback have been demonstrated. The cavity-locked laser is stabilized to a saturated absorption resonance in rubidium which is modulated and shifted by a combination of ac and dc magnetic fields. A linewidth of 15 kHz, long term stability of 10 kHz, and tunability over 80 MHz with no laser frequency modulation has

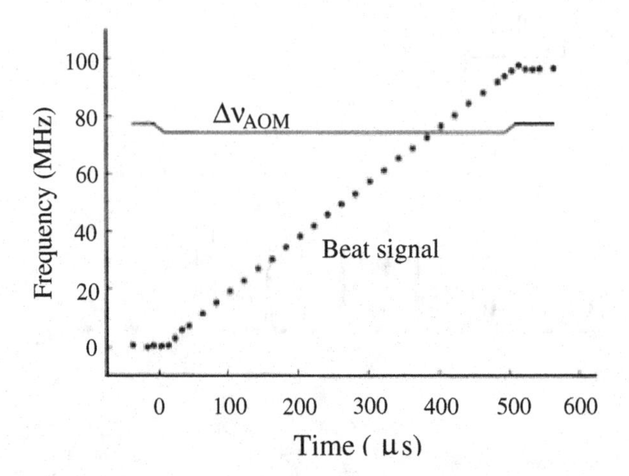

Fig. 7.15 (a) Laser frequency tuning by control of $\Delta\nu_{AOM}$. Circles, beat signal frequency versus time, solid line, $\Delta\nu_{AOM}$ versus time. Adapted with permission from *Opt. Lett.* **23**, 16, pp. 1280-1282. Greiner et al. (1998).

been achieved [Dinneen *et. al.* (1992)]. More recently, Hayasaka [Hayasaka (2002)] reported short term frequency stabilization of an external cavity diode laser at 420 nm by resonant optical feedback from a confocal cavity. A short term frequency stability of 300 kHz/s has been achieved and a resolution-limited linewidth of 880 kHz was measured. A new method has been used to keep the optical feedback phase and the external cavity diode laser frequency to the optimum values at same time, which is essential for continuous frequency scans over several GHz and for a stable lock for hours.

Chapter 8

Applications of Tunable External Cavity Diode Lasers

Having discussed how to develop and use the ECDL system with single mode, narrow linewidth, and wide tunability, we can now proceed to develop some applications in scientific research and engineering. A number of new applications of the novel capabilities of external cavity diode lasers have been demonstrated [Figger *et. al.* (2002); Ohtsu (1991)], Practical purposes that can be achieved by using a tunable external cavity diode laser system may be summarized as follows:

- Atomic clock and magnetometer
- Ultra-high resolution spectroscopy
- Quantum and nonlinear optics
- Quantum manipulation and engineering
- Actively mode-locked diode lasers
- Optical coherent communication system and lidar technology
- Gas monitoring sensor

8.1 Atomic clocks and magnetometry

All clocks are composed of two major parts, one generating periodic events, the other counting, accumulating and displaying these events. However, an atomic clock needs a third component, the resonance of a well-isolated atomic transition used to control the oscillator frequency. If the frequency of the oscillator is made to match the transition frequency between two non-degenerate atomic levels, then the clock will have improved long-term stability and accuracy. For an atomic clock based on a microwave transition, high-speed electronic count and accumulate an integer number of cycles of the reference oscillator to make a unit of time. The same basic concepts apply for a atomic magnetometer, which has been around for more than 40 years, and exploits the effects of magnetic fields on the unpaired spin of individual atoms.

8.1.1 *Atomic clock*

One of the earliest applications of a diode laser to atomic physics was used as a highly coherent light source for optical pumping and probing of a polarized medium. Many applications of this type has been reviewed by Camparo [Camparo (1985)]. Here we introduce the application of atomic clocks via optical pumping by use of diode lasers. Diode lasers have long been used extensively in frequency standards and atomic clocks, in these applications they optically pump the atoms of interest into specific level, then detect the atoms that have undergone the microwave frequency clock transitions [Vanier and Audoin (1989)]. Fortunately, the atoms used in some of the best frequency standards (rubidium and cesium) have resonance line wavelengths that are easily generated by semiconductor lasers.

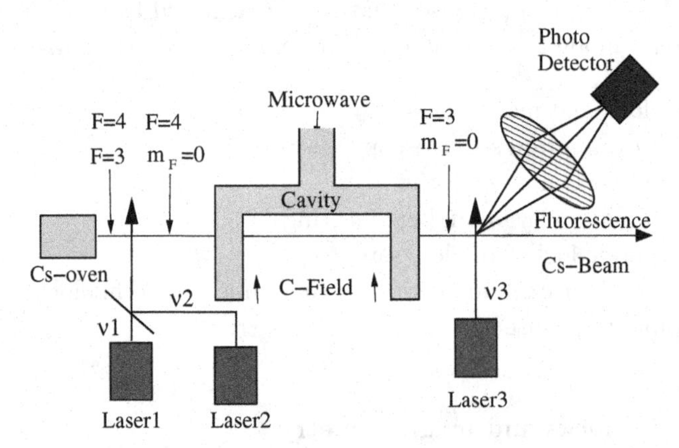

Fig. 8.1 Diagram of an optically pumped cesium atomic of frequency standard. The cesium atomic beam travels in a vacuum through the first laser interaction region where the atoms are optically pumped, it then traverses the microwave cavity region and then on to the second interaction region where the atoms are detected. The population in the ground state hyperfine levels is indicated along the path by the F and m_F values. The microwave clock transition is between the F=4, $m_F = 0$ and to F=3, $m_F = 0$ states. As shown the system used two lasers with different frequencies for optical pumping and a third laser for detection. Adapted with permission from *Rev. Sci Instrum.* **62**, 1, pp. 1-20. Wieman and Hollberg (1991).

Figure 8.1 is a schematic diagram of a diode-laser pumped cesium atomic clock. The lasers are first to prepare the atomic beam by placing all atoms into the specific hyperfine level using optical pumping. Then, after the atoms have passed through the microwave region, a laser excites the initially depleted level. Thus the microwave transitions are detected by the increase of the downstream fluorescence.

Alternative techniques such as optically pumped microwave double resonance and Raman scattering have been investigated over years to provide more compact and low-cost frequency reference for applications [Poelker *et. al.* (1991); Vukicevic *et. al.* (2000)]. A small atomic clock based on coherent population trapping resonances has been reported in a thermal cesium atomic vapor. A single-mode VCSEL is used as coherent radiation source at wavelength 852 nm with a linewidth of 50 MHz and power of 987 μW. Laser beam becomes circular polarized light after passing through $\lambda/4$ wave plate. When the VCSEL's injection current is modulated at 4.6 GHz, half of the ground state hyperfine splitting of cesium, the two first-order side bands can be used to excite the atomic resonance and trap the atoms in the coherent superposition of the two ground dark states [Alzetta *et. al.* (1976); Arimondo (1996)].

The laser beam is shining on the Cs vapor cell and a photo detector collects the transmitted light, the laser frequency is tuned in such a way that the two first-order side bands are resonant with the transitions from the two 6 $S_{1/2}$ ground state components to the excited 6 $P_{3/2}$ state. When the 4.6 GHz source is tuned over the dark resonance, a dispersive error signal is produced with FWHM of 100 Hz. In this way the frequency of the atomic clock can be measured as a function of time with optimal experimental parameters [Kitching *et. al.* (2000); Kitching *et. al.* (2001)]. The clock performance instability drops at a rate of $9.3\times10^{-12}/\sqrt{\tau/s}$, comparable to good commercial rubidium frequency standards, where τ is average times.

8.1.2 *Atomic magnetometer*

Atomic magnetometer in principle works in similar way to the atomic clocks. As a vapor cell (Rb, Cs) is placed in the homogeneous magnetic field, the degenerated atomic levels are split into Zeeman sublevels by the external magnetic field. When circularly polarized light illuminates on the vapor cell, the atomic population of Zeeman sublevels becomes substantially different from those obeying Boltzmann thermal distribution. Due to the selection rule for circularly polarized light $\Delta m_F = \pm 1$, eventually, most of population will pile up at certain Zeeman sublevels with maximum m_F.

The ensemble of atoms has been highly polarized by means of optical pump, and there is no more absorption of light when the equilibrium is reached between the optical pumping and all the relaxation processes in this atomic system. At this point, if the transverse radio frequency (RF) radiation field resonant with the transition of Zeeman sublevels is applied, one can observe a sudden decrease of transmission of applied light and consequently measure the magnitude of magnetic field strength from the resonant RF frequency.

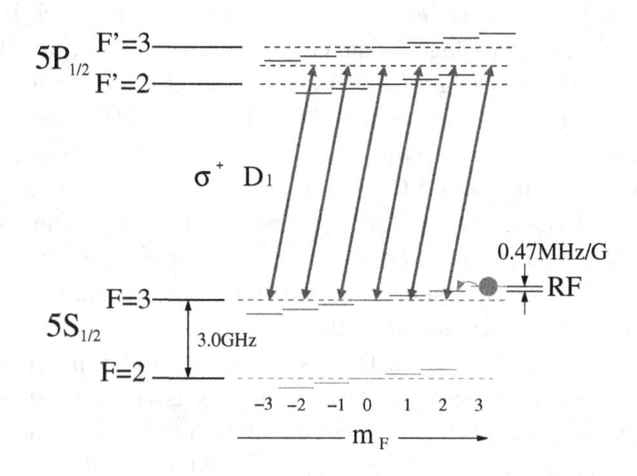

Fig. 8.2 Schematic of energy levels, optical pumping with radio frequency radiation and double resonance in Rb.

Many different styles of atomic magnetometers have been developed [Fitzgerald (2003)]. One approach [Alexksandrov *et. al.* (1996)], pioneered by Alexander, is a double resonance technique that combined optical pumping with radio frequency radiation, as shown in Fig 8.2. Such magnetometers have attained sensitivity of 1 $\text{fT}/\sqrt{Hz}$. Recently, a new type of magnetometer based on electromagnetically induced transparency (EIT) [Harris (1992)] was proposed by Scully and Fleischhauer [Scully and Fleischhauer (1992); Fleischhauer and Scully (1994)], and experimentally demonstrated [Nagel *et. al.* (1998); Affolderbach *et. al.* (2002)]. Two lasers are used to excite the transitions from different hyperfine levels of ground states to a common excited state like the case of atomic clock based on coherent population trapping. When the lasers are tuned to be in two-photon Raman resonance, they drive population into so-called coherent dark state that decouple from the laser radiation. The transparent otherwise opaque laser beam characterizes narrow line width, and thus provides a highly sensitive detection of magnetic field. This technique has achieved sensitivities around 1 pT/$\sqrt{Hz}$. In another approach, developed by Budker et al. and Novikova et al. [Budker *et. al.* (2000); Novikova (2001)]. The nonlinear Faraday effect has been employed to determine the magnetic field with a theoretical sensitivity limit of 0.3 fT/$\sqrt{Hz}$ [Budker *et. al.* (2002)].

8.2 High resolution laser spectroscopy

High-resolution laser spectroscopy could provide accurate information on the structures and dynamic properties of the excited states of atoms and molecules. Tunable ECDLs are ideally candidate as a coherent radiation sources for high resolution spectroscopy of atoms and molecules. The advantage of spectroscopy by use of ECDLs is the possibility of obtaining a wide band frequency sweep by varying the temperature or injection current of diode lasers as well as external cavity tuning. In this section, we show some applications of ECDLs in atomic and molecular spectroscopy by describing experiments recently performed in the violet-blue (390$\sim$ 420 nm), the visible (650$\sim$ 690 nm), and near infrared (750$\sim$ 895 nm). We discuss the spectral resolution, the accuracy of frequency measurements, and the detection sensitivity achievable with ECDLs.

Many examples of high resolution spectroscopy have been described [Tanner and Wieman (1988); Wieman and Hollberg (1991); Inguscio and Wallenstain (1993); Inguscio (1994); Bhatia *et. al.* (2001)]. A myriad of works have been done in the wavelength region around 780 nm and 850 nm, but only a limited number of experiments have been reported around 400 nm and 670 nm. For all spectroscopic applications a wide mode hop-free tunability of the laser output frequency is desirable. A diode laser with an external grating can be tuned by several means: diode laser temperature, diode injection current, grating distance, and grating angle. The tuning range can be expanded by employing a feed-forward loop: an attenuated version of the grating piezo voltage is added to DC injection current of the diode. Fine adjustment of the attenuation factor and DC operating point enable the mode hop-free scan range to expand from 6 GHz to 20 GHz. Doppler-free spectroscopy on an indium atomic beam by blue-emitting diode laser by use of GaN diode laser in the Littrow configuration has been reported [Leinen *et. al.* (2000);

Hildebrandt *et. al.* (2003)].

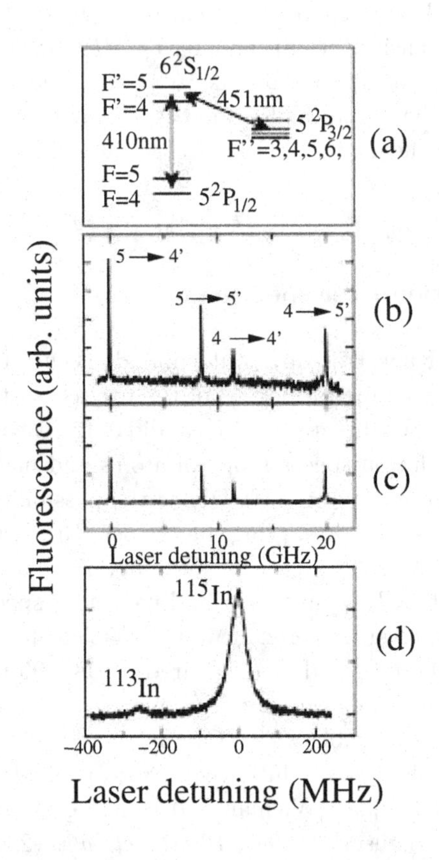

Fig. 8.3 (a)Schematic of energy level of Indium concerned. (b) Hyperfine resolved total fluorescence spectrum of the transition $5\ ^2P_{1/2} \rightarrow 6\ ^2S_{1/2}$ in an indium atomic beam. (c) Spectrum take with the bandpass filter for 450 nm. (d) The lowest-frequency peak measured with better resolutions: the contribution of the ^{113}In isotope becomes noticeable. Adapted with permission from *Appl. Phys. B* **70**, pp. 569. Leinen et al. (2000).

The fluorescence spectrum of indium atoms in an atomic beam by absorption of 410 nm laser light is shown in Fig. 8.3. The relevant energy level diagram is shown in Fig. 8.3(a), the $6\ ^2S_{1/2}$ excited state of indium can be reached from $5\ ^2P_{1/2}$ ground state by GaN diode laser, it can decay into either the $5\ ^2P_{3/2}$ or back into the $5\ ^2P_{1/2}$ state with emission wavelength of 451 nm or 410 nm light, respectively. Fig. 8.3(b) shows the fluorescence spectrum collected by a photomultiplier tube (PMT). All four hyperfine components for transitions from F=4, 5 to F'=4, 5 are resolved within a single mode hop-free scan. Fig. 8.3(c) shows much less noisy spectrum of Fig. 8.3(b) after using a narrow-band interference filter center at 450 nm to block the residual stray laser light. One can distinguish the contributions from the two isotopes ^{113}In and ^{115}In in the higher resolution fluorescence spectrum, each fluorescence peak in Fig. 8.3(b) and (c) is actually a doublet as resolved in Fig. 8.3(d), their relative height of the two peaks corresponds to the natural abundance ratio of the two isotopes.

Very recently Hayasaka [Hayasaka (2002)] has performed saturation spectroscopy in the $5S_{1/2} \rightarrow 6P_{3/2}$ rubidium transition using the frequency stabilized violet diode laser at wavelength 420 nm in a gas cell, and observed the Doppler-free spectra with a linewidth of 2.3 MHz as shown in Fig. 8.4 (a) and (b), four groups of hyperfine transition in ^{87}Rb and ^{85}Rb are resolved. Fig. 8.4(a) illustrates two groups of the transition recorded by a continuous frequency scan over 2.8 GHz. The signals on the left are hyperfine transition from ^{87}Rb, F=2 and those on the right are from ^{85}Rb. An enlarged view of the transitions located in the left is displayed in Fig. 8.4(b), in the figure the total angular moment of upper state is denoted by $F^{'}$. All the components including crossovers have been observed. The same group of Hayasaka's [Uetake (2002)] reported the saturation spectra of potassium of the $4S_{1/2} \rightarrow 5P_{1/2}$ transition with a linewidth of 4 MHz by using feedback stabilized violet diode laser at wavelength 404.8 nm with similar experimental setup as shown above, the spectrum is shown in Fig. 8.5. To obtain the spectra, the pump beam was chopped at 2 kHz by an optical chopper and the probe beam was detected by a lock-in amplifier.

In region of visible light, we give two examples to illustrate the application of ECDLs in atomic spectroscopy. In the work of Hof et al. [Hof *et. al.* (1996)], Littrow configuration of red diode laser was used to induce the fluorescence from an atomic beam of ^{6}Li in a magnetic field of 53 mT. The Zeeman sublevels of D_1 transition with a upper limited laser linewidth of < 25 MHz was measured. Fluorescence spectrum has previously been measured by Boshier et al. in the transition of $2S_{1/2} \rightarrow 2P_{1/2}$ of lithium at 670.8 nm [Boshier *et. al.* (1991)]. A saturated absorption spectra of lithium (with natural abundance) has been demonstrated by Libbrecht et al. [Libbrecht *et. al.* (1995)], by use of the stabilized 670 nm diode lasers in undergraduate teaching laboratory. Doppler-free peak from both ^{7}Li and ^{6}Li superposed on the Doppler-broadened absorption profile has been observed.

Cesiums D_2 transition at 852 nm is easily reached by diode laser. There

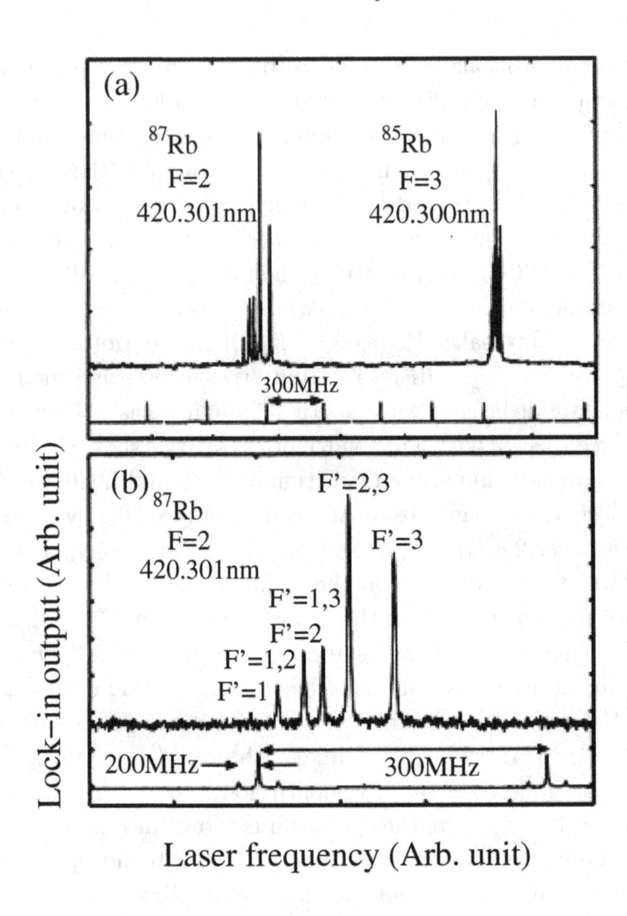

Fig. 8.4 Doppler-free spectra of $5^2S_{1/2} \rightarrow 6^2P_{1/2}$ transition in Rb at 420 nm. (a) Spectra taken at a continuous frequency scan of 2.8 GHz. (b) Magnification of the spectra of the left-hand side of (a). F and $F^{'}$ denote the total angular momentum of the ground state and that of the upper state, respectively. Adapted with permission from *Opt. Commun.* **206**, pp. 407. Hayasaka (2002).

is no routinely available diode laser at 894 nm D_1 transition. To drive this transition, one has to rely on severely expensive and complicated titanium-sapphire or dye laser. A simple Doppler-free spectroscopy experiment on the four hyperfine transition of the D_1 line has been demonstrated [Ross *et. al.* (1995)]. Two equal power counter propagating laser beams from ECDLs were overlapped through the cesium cell. The laser was scanned across the D_1 transition. The result is shown in Fig. 8.6. As expected, there are two sets of two lines, the pairs separated from each other by

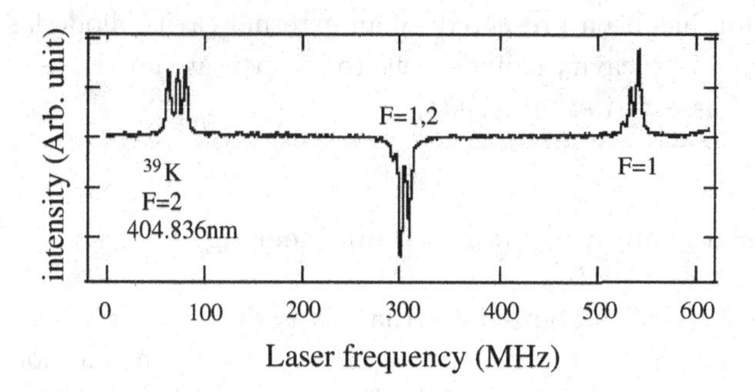

Fig. 8.5 Saturation spectra of potassium. Adapted with permission from *XVIII International Conference on Atomic Physics,* Cambridge, Massachusetts, USA, pp. 43.

ground state hyperfine splitting of 9.193 GHz, the splitting in both pairs corresponds to the upper state hyperfine splitting 1.168 GHz. A Lamb dip is present in each line.

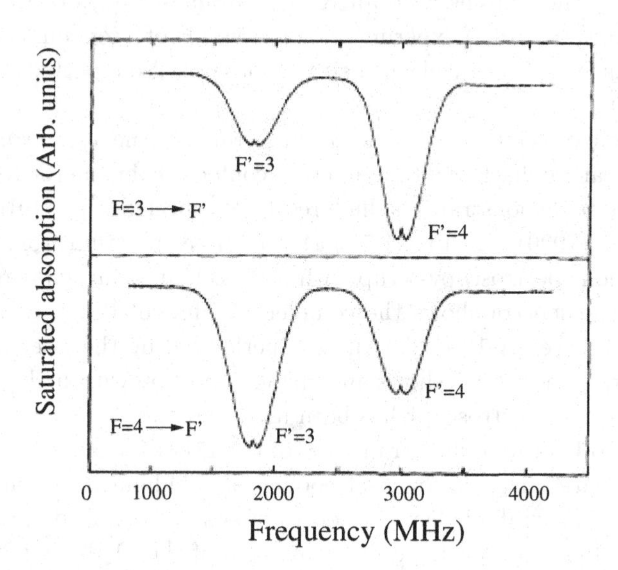

Fig. 8.6 The four hyperfine $6^2S_{1/2}(F) \rightarrow 6^2P_{1/2}(F')$ recorded by saturation absorption spectroscopy. The Lamb dip exists in each line. The Doppler width of $F = 3 \rightarrow F' = 4$ transition is 380 MHz; the linewidth of the saturated peak is 40 ± 5 MHz. Adapted with permission from *Opt. Commun.* **120**, pp. 156. Ross et al. (1995).

A saturated-absorption spectroscopy of the $D1$ line of cesium at 895 nm with Lamb dip has been presented by an external cavity diode laser characterized by an intracavity cylinder lens to compensate for the astigmatism of the laser [Cassettari *et. al.* (1998)].

8.3 Quantum manipulation and engineering

Another area in which stabilized external cavity diode lasers have been used very successfully is in cooling and trapping neutral atoms and ions. Laser cooling was first proposed in 1975 by Hansch and Schawlow [Hansch and Schawlow (1975)], and simultaneously by Wineland and Dehmelt [Wineland and Dehmelt (1975)]. Weidemuller et al. [Weidemuller *et. al.* (1993)] have achieved cooling down lithium atom into Rydberg states by using diode laser for high-resolution of microwave spectroscopy. Their velocity can be efficiently decreased and controlled by periodically chirping the frequency of the first step excitation diode laser at a repetition rate of 500 ps. During each chirp sequence, atoms with all velocity from 1500 m/s down to zero are successfully prepared. Experimental observation of laser confinement of neutral atoms in optical-wavelength-size regions was also reported [Salomon *et. al.* (1987)].

A narrow bandwidth tunable semiconductor laser near 780 nm stabilized by optical feedback from an external confocal Fabry-Perot resonator has been used to demonstrate the high resolution spectroscopy of Rb [Hemmerich *et. al.* (1990)]. In Fig. 8.7, a Doppler-free spectrum from a standard saturation spectroscopy setup with a Rb cell at room temperature is presented. The spectra shows the complete D_2 line of Rb, with resolved hyperfine structure. A 10 GHz scan was performed by the laser in 5 ms. The upper trace in Fig. 8.7 shows an expanded portion with higher resolution. Polarization spectroscopy has been used to obtain a dispersive signal for the F=3 to F=4 hyperfine transition of the D_2 of ^{85}Rb.

The schematic diagram for laser cooling of rubidium is shown in Fig. 8.8. Three stabilized diode laser systems were employed. Two of them are tuned to the F=3 to F=4 hyperfine transition of the D_2 line of ^{85}Rb for cooling and velocity probe. The third laser is tuned to the F=2 to F=3 transition to repute atoms which are optical pumped away from the cooling transition. The cooling and pumping laser counter-propagating against the atomic beam while the probe laser intersects the atomic beam at a 45 ^{o} angle below a photomultiplier fluorescence detector. The laser frequency are synchronously scanned by two time gates. Plots of manipulated Doppler

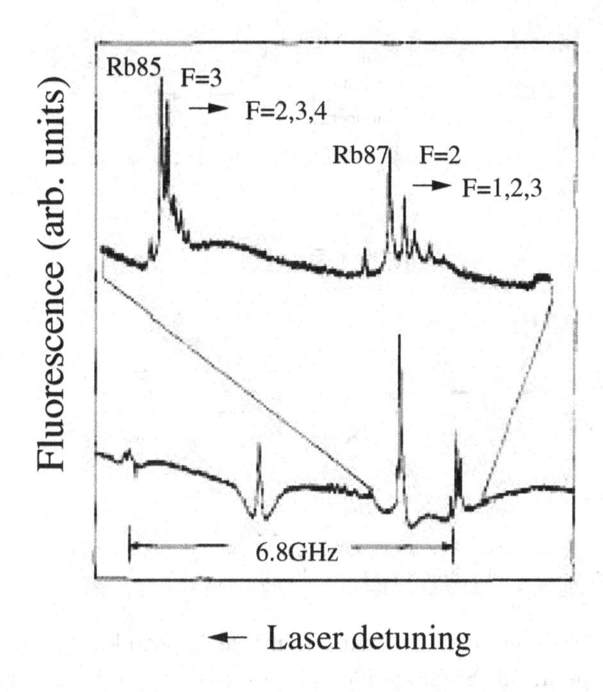

Fig. 8.7 Doppler-free saturation spectra of the D_2 line of rubidium demonstrating the scanning ability of the laser system. The upper traces shows an expanded portion taken with a scan speed of 100 MHz/ms. The three hyperfine lines and the three corresponding crossover lines are resolved. Adapted with permission from *Opt. Commun.* **75**, 2, pp. 121. Hemmerich et al. (1990).

profiles are shown in Fig. 8.9, which demonstrate both positive and negative final velocities.

Myatt et al. [Myatt *et. al.* (1993)] have demonstrated direct microwave modulation of diode lasers operated with optical feedback from a diffraction grating, they obtained substantial fractions of the laser power (2-30%) in a single sideband at frequencies as high as 6.8 GHz with 20 mW of microwave power and simple inefficient microwave coupling. Using a single diode laser modulated at 6.6 GHz, they trapped ^{87}Rb atoms in a vapor cell. With only 10 mW of microwave power they trapped 85 % as many atoms as were obtained by using two lasers in the conventional manner. Calcium ions are laser cooled on Paul trap by use of a grating stabilized UV diode laser, and ions are successfully cooled to crystallization temperature with sufficient reproducibility [Toyoda *et. al.* (2001)].

A UV external cavity diode laser is used as the cooling laser in the

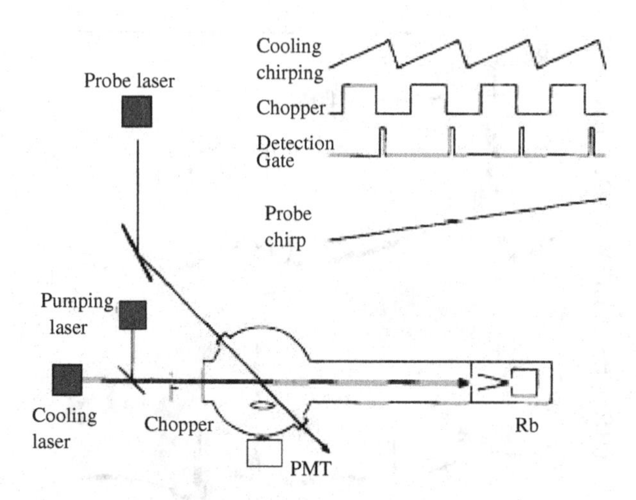

Fig. 8.8 Experimental setup for laser cooling. Adapted with permission from *Opt. Commun.* **75**, 2, pp. 121. Hemmerich et al. (1990).

experiment, this laser oscillates with multiple longitudinal modes in the free running operation. Single-mode operation is obtained by construction of an external cavity with a Littrow configuration with holographic grating and a collimating lens. The output power of the external cavity laser is 1.9 mW at an operating current of 41 mA. The output beam is sent through an optical isolator, then it is divided into two beams, one for cooling, the other for monitoring the single mode during scanning.

8.4 Actively mode locked diode lasers

There are many applications of very short-duration laser pulses in optical communications. The technique that has allowed the generation of optical pulses as short as 6 fs is referred to as mode-locking, which can be achieved by combining a number of distinct longitudinal modes of a laser, all having slightly different frequencies and some definite relation between their phases [Siegman (1986)]. Mode locking in multimode diode laser operation is a well-know phenomenon in which mode separations near each other and relative phases between oscillating modes are fixed so that beat notes are emitted from the diode laser. The pulse separation corresponds to the time it takes for an optical pulse to complete one round trip in the cavity. There are scientific and engineering interests in the mode locking

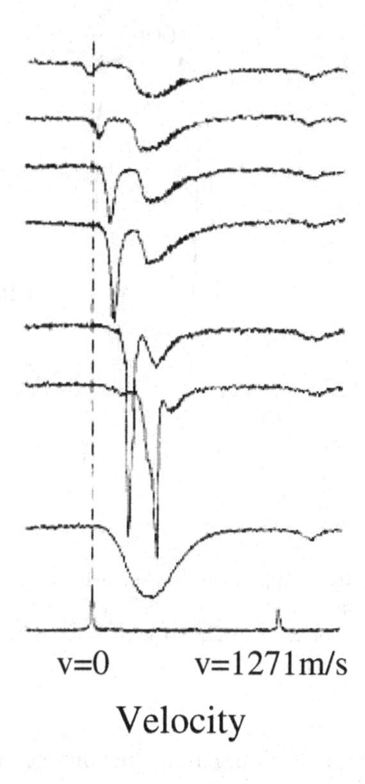

Fig. 8.9 Doppler-profiles of the laser cooled rubidium beam. The top trace shows atoms cooled to a negative final velocity, the next five traces show atoms cooled to increasing positive final velocity. The significant changes in the cooled peak heights are caused by the limited length of the atomic beam machine. Also shown at the bottom are Doppler profile of the non-cooled beam and frequency markers obtained from perpendicular excitation. The first marker determines zero velocity and the second marker (the F=2 to F=3 ^{87}Rb) allows one to determine the velocity scale. Adapted with permission from *Opt. Commun.* **75**, 2, pp. 122. Hemmerich et al. (1990).

of the semiconductor lasers, because a mode locked laser can emit very short optical pulses with almost ideal time-bandwidth products and high repetition rates [Haus (1980); Kuhl *et. al.* (1987); Schell *et. al.* (1991); Delfyett *et. al.* (1992)]. Thus it finds a wide variety of applications in electro-optic sampling, time-division multiplexing, fiber optic communications [Liu *et. al.* (1992); Goldberg and Kliner (1995); Yilmaz (2002)], to name a few. The successful operation of an actively mode locked GaAlAs laser stimulated a great deal of research aimed at the generation of picosecond optical pulses in semiconductor diode laser devices [Ho *et. al.* (1978);

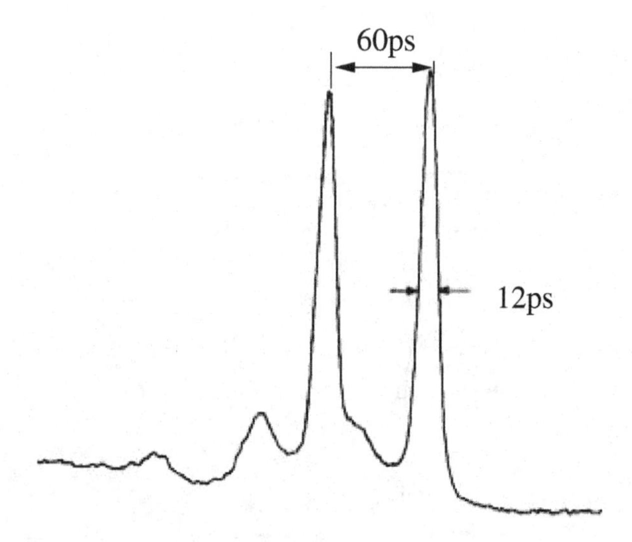

Fig. 8.10 Intensity profile of streak image showing calibrated delay of 60 ps and recorded pulse duration of 12 ps. Devolution of synchronous scan streak camera resolution gives the actual laser pulse duration of 9 ps. Adapted with permission from *Opt. Commun.* **48**, 6, pp. 429. Chen et al. (1984).

Holbrook *et. al.* (1980)], conventional mode locking techniques or excitation of semiconductor laser diodes by short electrical pulses have been applied. A brewster-angled GaAlAs semiconductor diode laser has been actively mode-locked in an external cavity Littrow configurations to provide tunability over 15 nm and pulse duration $\sim$ 9.5 ps, high stability of spectral output and pulse duration was observed [Chen *et. al.* (1984)] as show in Fig. 8.10. A simple technique has been developed to produce ultrashort, coherent, wavelength-tunable pulses from gain-switched Fabry-Perot lasers [Cavelier *et. al.* (1992)]. Pulse durations of 2.5 ps have been obtained at 1.3 μm with a time-bandwidth product of 0.6, a repetition rate adjustable up to 12 GHz and a wavelength tunability of 20 nm, the pulse energy is around 1 pJ. A schematic of experimental setup is shown in Fig. 8.11, it consists of a gain-switched FTP laser in a long (3.5 cm) highly attenuating and selective external cavity. The end reflector of the cavity is a 1200 g/mm grating. An optical attenuator limits the feedback to the laser diode to extract 30 % of the laser beam. Gain switching of the laser diode is realized by strong sinusoidal current modulation. Two 1 W microwave amplifiers were used to cover the frequency range from 0.7 GHz to 12 GHz. Figure 8.12

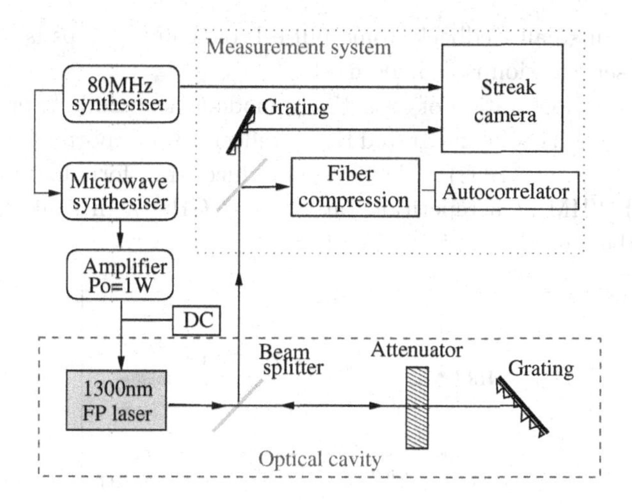

Fig. 8.11 Schematic of experimental setup. Adapted with permission from *Electron. Lett.* **28**, 3, pp. 224. Cavelier et al. (1992).

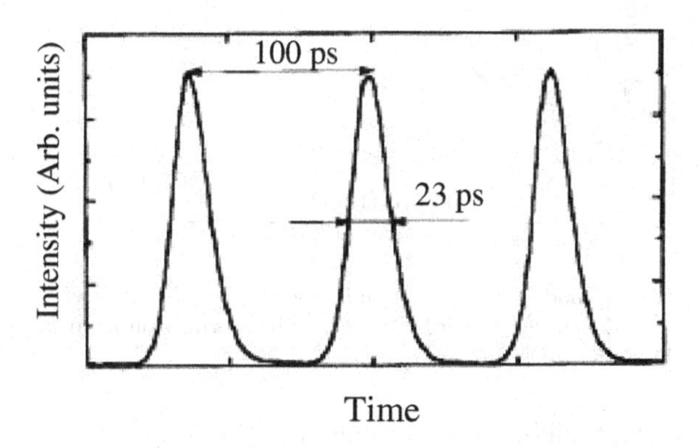

Fig. 8.12 Laser pulse train measured by streak camera with 10 GHz modulation frequency. Adapted with permission from *Electron. Lett.* **28**, 3, pp. 225. Cavelier et al. (1992).

shows a typical train of gain-switched multi-mode laser pulse measured by the streak camera when there is no optical feedback. The laser is operated at 10 GHz repetition rate with DC bias of 100 mA and the amplitude of the modulation current of 200 mA. The measure pulse width is 23 ps which includes the 10 ps time resolution of the camera and some weak additional jitter. Separate measurement by autocorrelation indicates an actual width of 14 ps. With small feedback, some pulse broadening $\sim$ 1 ps is observed when the laser emission is a single mode.

The first demonstration of an actively mode-locked diode laser using a fiber external cavity with integrated Bragg reflector was reported by Morton et al. [Morton *et. al.* (1992)]. The device produces transform limited pulses of 18.5 ps FWHM at a repetition rate of 2.37 GHz, high peak power of 49 mW in the output fiber is achieved.

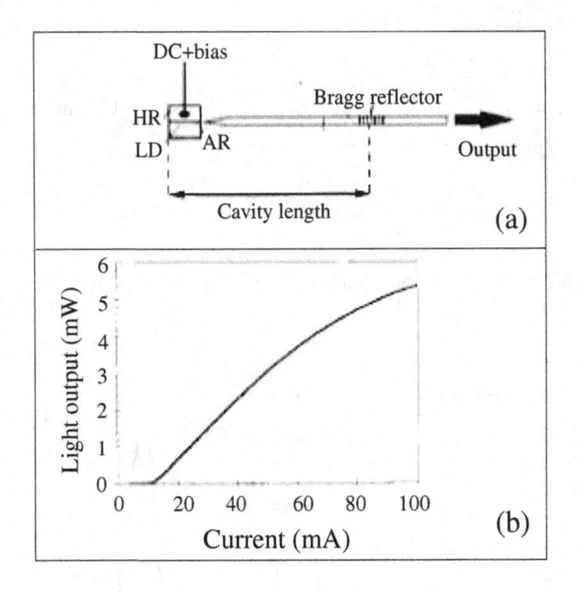

Fig. 8.13 Schematic of mode-locked laser and typical DC light feature. (a) Schematic diagram. (b) Typical L-I characteristics. Adapted with permission from *Electron. Lett.* **28**, 6, pp. 561. Morton et al. (1992).

A schematic diagram of the experimental setup is shown in Fig. 8.13(a). A 1.55 μm laser diode is used with one facet high reflectivity coated for improved cavity quality factor Q, and the other antireflection coated to allow coupling to the external cavity and suppress Fabry-Perot modes. The external cavity is composed of an AR coated lenses fiber stub fusion spliced to a section of photosensitive fiber with an integrated Bragg reflector. This arrangement allows efficient coupling from the laser diode to the fiber. The grating has an approximately Gaussian taper with a 5 mm FWHM, giving rise to a Bragg reflector with a reflectivity FWHM of 0.2 nm.

The effective cavity length of the device depends on the position of the fiber grating and the operating wavelength. A DC bias and microwave drive are applied to the laser diode, with modulation period being close to the fundamental mode locking frequency. In the results presented here, a step recovery diode is used to provide short (40~50 ps) electrical pulses to the laser diode at the mode-locked frequency. The output of the device is passed through an optical isolator, and then split between measurement instruments including a 50 GHz sampling oscilloscope with bandwidth of 32 GHz detector, an optical spectrum analyzes, average power meter and a bandwidth of 22 GHz microwave spectrum analyzer.

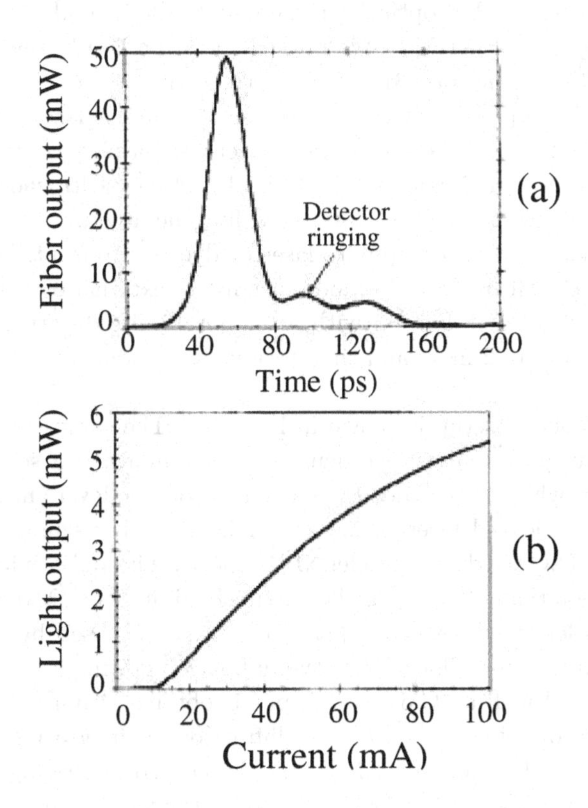

Fig. 8.14 Measured optical pulse shape and optical spectrum under optimum mode-locked condition at the repetition rate of 2.37 GHz. (a) Optical pulse shape, (b) L-I characteristics curve. Adapted with permission from *Electron. Lett.* **28**, 6, pp. 562. Morton et al. (1992).

The DC light current characteristics is shown in Fig. 8.13(b). The threshold is low (11 mA) and the output power comparable to a standard laser diode (>5 mW) at (100 mA). The optimum mode-locking conditions are found to be a DC bias current of 36 mA and an RF frequency of 2.37 GHz with approximately +27 dBm power into the step recovery diode. The measured optical pulse shape is shown in Fig. 8.14(a), as seen on the sampling oscilloscope, with a measured pulse width FWHM of 24 ps. When decontrolled with a simple sum of square formula, assuming an oscilloscope FWHM of 8 ps and a detector FWHM of 13 ps, this gives the actual pulse width of 18.5 ps. The peak power level is calculated from the average power measured under mode-locked conditions which is 21 mW, giving a peak power of 49 mW. The optical spectrum under these condition is shown in Fig. 8.14(b). The FWHM of the optical spectrum is 0.13 nm, giving a time-bandwidth product of 0.31, which is transform limited.

With the developments of GaN-based semiconductor lasers, it is possible for semiconductor lasers to cover the entire wavelength range from ultraviolet to infrared [Nakamura and Fasol (1997)]. Ultrashort optical pulses of wavelength near 400 nm are typically generated by the frequency doubling of mode-locked Ti:sapphire laser or dye laser output, but it can be generated as well by actively mode locking an external cavity INCAN laser at a wavelength of 409 nm with a temporal pulse duration of 30 ps and average power of 2 mW, and time-bandwidth product of 1.2 [Gee and Bowers (2001)].

The experimental setup is shown in Fig. 8.15. The cavity is composed of a diffraction grating of groove density 1200 g/mm and a semiconductor diode laser, which is obtained by $SIC_2Ta_2O_5$ double-layer antireflection (AR) coating a cleaved facet of a commercial ridge waveguide multiple-quantum-well INCAN diode (model:NLHV500A). The reflectivity is estimated to be less than 10^{-3} using the Hakim-Pauli method [Merritt *et. al.* (1995)]. In order to suppress the Fabry-Perot mode caused by imperfect AR coating on the laser diode facet, the diffraction grating is employed to limit the spectral width. The laser output is obtained from a 50% beam splitter instead of using the zero-order diffraction of the grating owing to its low efficiency. The active mode locking is realized by driving the laser diode with 20 mW of RF power at the cavity round-trip frequency combined with 43 mA of DC current using a bias tee and with the repetition rate of 720 MHz and the average output power of 2 mW.

Figure 8.16(a) shows the optical intensity spectrum of the mode-locked laser output showing a spectral width of 0.023 nm at the center wavelength of 408.5 nm. Due to the imperfect AR coating on the diode laser, Fabry-Perot mode still exists with 0.035 nm of modulation period. Fig. 8.16(b) shows the temporal intensity profile of the output pulse measured by a

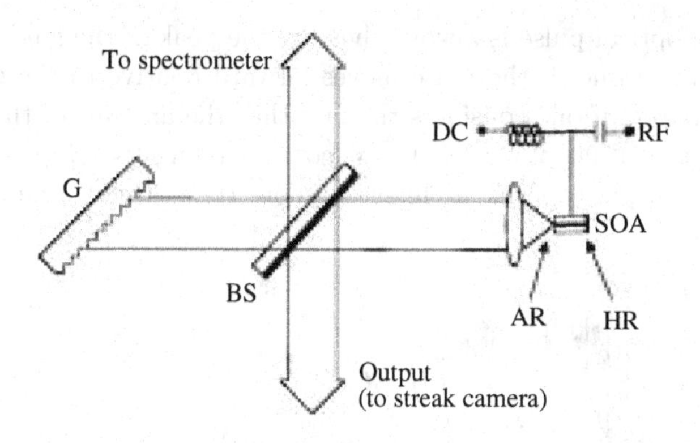

Fig. 8.15 Schematic of laser setup. SD: semiconductor diode laser, BS: beam splitter. Adapted with permission from *Appl. Phys. Lett.* **79**, 13, pp. 1951. Gee and Bowers (2001).

sacrosanct streak camera with resolution of 10 ps. The pulse has a steep rise and slow decay with the temporal width of 30 ps, indicating a time-bandwidth product (TBSP) of 1.2.

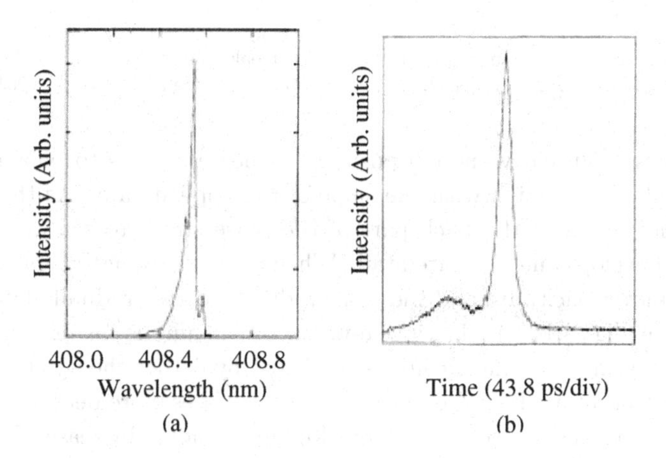

Fig. 8.16 Typical mode-locked output: (a) optical intensity spectrum and (b) streak camera images. Adapted with permission from *Appl. Phys. Lett.* **79**, 13, pp. 1952. Gee and Bowers (2001).

Figure 8.17 is the series of steak camera images for different cavity length setting while the RF bias current frequency is held constant at 720 MHz. The shadowed curve is the bias current wave form, which was determined by measuring the spontaneous emission of the diode laser. It can be easily seen that the optical pulse is always ahead of the peak of the gain. As the cavity length is reduced, the pulse moves forward relative to the gain. It can be understood if one considers the fact that the amount of the cavity length change in each plot is 250 μm, corresponding to 1.7 ps of delay while the actual pulse position change is more than 20 ps. The shortest

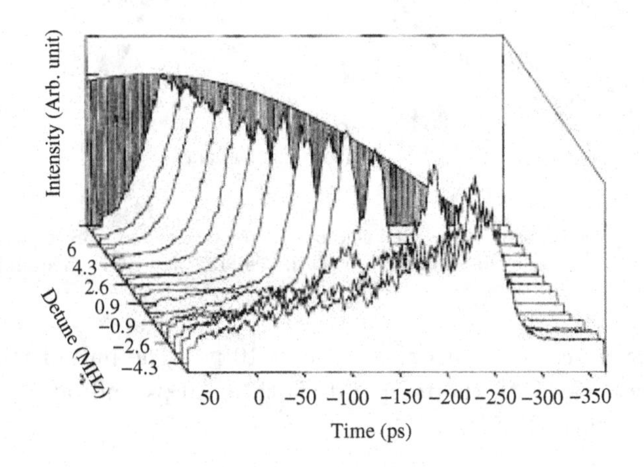

Fig. 8.17 Cavity length detuning characteristics: streak camera vs detuning. Adapted with permission from *Appl. Phys. Lett.* **79**, 13, pp. 1952. Gee and Bowers (2001).

pulse occurs at an early enough position of the gain so as to balance the higher level of saturation with the slope of the applied gain. As the pulse goes further forward, the back part of the pulse starts to see more gain and pulse develops and elongated tail. When the cavity length is increased, the pulse moves backward and the pulse width becomes gradually longer as indicated in Fig. 8.18(a). The pulse-width broadening is faster for cavity length decreasing than increasing as predicted by earlier theory. It should also be noted that as the pulse moves backward it sees more net gain due to the applied gain modulation and it resulted in a gradual increase of output power [Fig. 8.18(b)].

Figure 8.19 shows the RF bias current dependence of the pulse width. The pulse width decreases as the RF bias increases, showing that the pulse width is inversely proportional to the fourth root of the modulation strength while the time-bandwidth product almost remains constant (the dotted line is the fit for the negative fourth root of the RF bias).

An active mode locking has recently been reported for a 1.55 μm semiconductor laser with a curved waveguide and passive mode expander, sit-

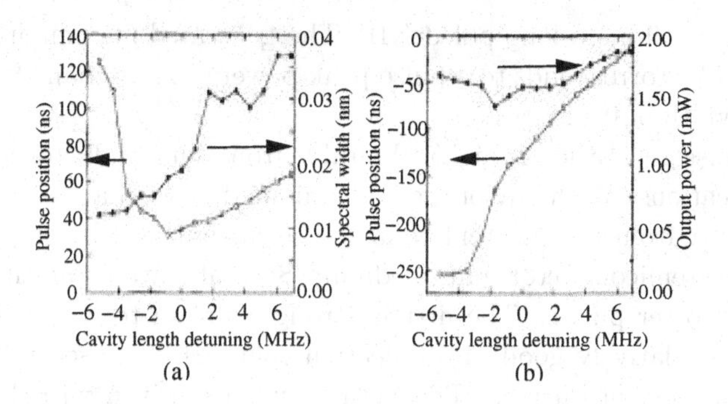

Fig. 8.18 Cavity length detuning characteristics: streak camera vs detuning: (a) pulse width and spectral width vs detuning and (b) pulse position and output power vs detuning. Adapted with permission from *Appl. Phys. Lett.* **79**, 13, pp. 1952. Gee and Bowers (2001).

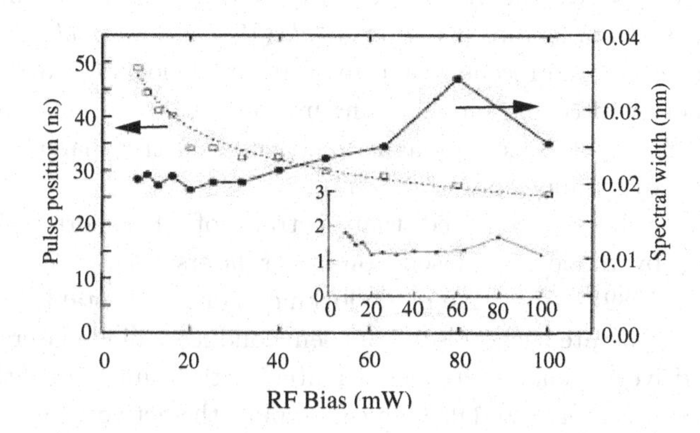

Fig. 8.19 RF bias characteristics: pulse width and spectral width vs RF bias. The inset is TBSP vs RF bias. Adapted with permission from *Appl. Phys. Lett.* **79**, 13, pp. 1952. Gee and Bowers (2001).

uated in a wavelength tunable external cavity. One facet with a very low reflectivity of 8×10^{-6} is achieved through a curved active region that tapers into an underlying passive waveguide, thus expanding the mode to give reduced divergence. 10 GHz pulses of 3.1 ps duration have been generated with a linewidth of 0.81 nm [Williamson *et. al.* (2003)]. Woll et al. [Woll *et. al.* (2002)] recently have reported on the generation of 250 mW of coherent 460 nm light by single-pass frequency doubling of the mode-locked picosecond pulses at a frequency repetition of 4.8 GHz and 16.5 ps pulse width emitted by an InGaAs diode master oscillator power amplifier in periodically 10-mom-long poled KIP. The generated maximum average power of 3.4 W corresponds to a pulse peak power of more than 30 W with a spectral width of 0.14 nm.

Short pulse generation is highly desirable from semiconductor lasers for many applications. With any of these technique described above, the pulse widths are far from the inherent limit ($\sim$ 50 fs) imposed by the spectral width of the semiconductor gain medium. Several features are necessary to achieve shorter pulses. The first is strong coupling to external cavity, this requires relatively good anti-reflection coatings. The second feature is good light-current curves. The lasers must be driven with large RF driver to achieve short pulses. Perhaps the most important factor is large modulation bandwidth. The final important feature is frequency stability of the microwave oscillator, which is driving the laser. In this way, short pulses were obtained with laser biased near threshold, and average output power was 0.5 mW with nearly transform-limited hyperbolic secant pulses with a pulse width of 0.58 ps [Corzine *et. al.* (1988); Bowers *et. al.*(1989)].

It has been reported recently that an active mode locking external cavity diode laser in Littman-Metcalf configuration can excite the atomic coherence, which leads to electromagnetically induced transparency in hot rubidium (^{85}Rb) atomic vapor [Ye (2004)].

It has long been recognized that a train of ultrashort pulse can be generated by mode locked semiconductor lasers [Ho *et. al.* (1978); Bowers *et. al.*(1989)]. An active mode locking is one well-established technique used to generate such pulses from semiconductor diode lasers by applying a RF drive current at a frequency matching the round trip of the laser cavity [Gee and Bowers (2001)]. One can exploit the actively mode locked diode laser to demonstrate an electromagnetically induced transparency in hot Rb atomic vapor. With a repetition frequency $f_{rep} = 506\ MHz$ (cavity round trip frequency) applied to diode laser in an external cavity of Littman-Metcalf configuration, a train of pulse with pulse width of $\sim$ 250 ps has been observed when the spacing of a multiple of the modes of a multimode diode laser was coincident with the spacing of the hyperfine states, As shown schematically in Fig. 8.20. For ^{85}Rb, that is $qf_{rep} = \Delta\nu_{hs} = 3.035\ GHz$, which corresponds to the integer $q = 6$. Any two optical modes resonant with hyperfine states renders transparent otherwise opaque. We observed this transmission in terms of beat note of the two resonant optical modes on the spectrum analyzer.

The experimental arrangement is shown schematically in Fig. 8.21. A 780 nm semiconductor laser is used as a coherent radiation source, the

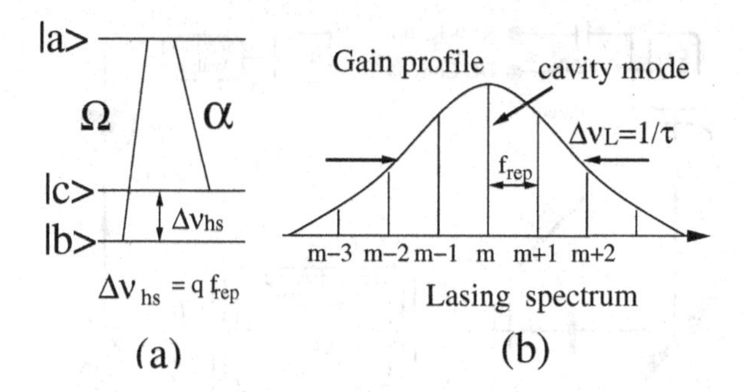

Fig. 8.20 (a) Three-level scheme for EIT. In our experiment, ground states $|b\rangle$ and $|c\rangle$ are hyperfine splitting of ground state of Rb $5S_{1/2}$ F=2 and F=3, respectively, $|a\rangle$ corresponds to the excited state $5P_{3/2}$. (b) Cavity modes in the gain profile of semiconductor diode laser.

residual reflectivity at the AR-coated facet is $\sim 0.01\%$. The laser diode is mounted on a thermoelectric-cooler to stabilize its temperature, it is situated in an external cavity formed between the conventional facet of diode laser and a moveable mirror. A diffraction grating of groove density 1200 g/mm at grazing incidence serves for wavelength selection and output coupling, the laser cavity is arranged in Littman-Metcalf configuration. The cavity round trip frequency can be adjusted by moving the external mirror. The external cavity length is approximately l=30 cm and the corresponding round-trip frequency is $f_{rep} = c/2l = 506\ MHz$, where c is the speed of light in vacuum. The active mode locking is realized by driving the laser diode at the cavity round trip frequency combined with 55 mA of dc current using a bias tee. The setup is mode locked in fundamental operation at repetition rate 506 MHz and the average output power is 500 μW.

The EIT was observed in ^{85}Rb vapor cell by monitoring the mode locked laser transmission through the cell when the repetition frequency is tuned around resonance frequency of 506 MHz and the laser frequency is set to the resonant transition of D_2 line of ^{85}Rb $5S_{1/2} \rightarrow 5P_{3/2}$ (780 nm) . Rb cell containing the natural abundance ^{87}Rb and ^{85}Rb is placed in the magnetic shielding cylinder chamber which screens the residual magnetic field. Cell's temperature can be controlled by electric heating tape which are designed to generate no extra magnetic field inside chamber, temperature is set around 70 $°C$ yielding an atomic density of $6 \times 10^{11} cm^{-3}$. The mode locked laser transmits through the half-wave and quarter-wave plates, respectively, becoming circular polarized light, then focuses into the cell. The transmitted laser is detected by the 25 GHz fast photodiode, the beat note

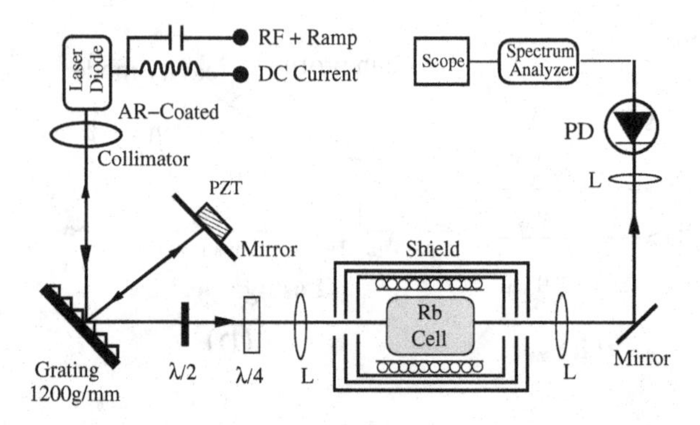

Fig. 8.21 Experimental setup. PZT: piezoelectrical transducer. $\lambda/2, \lambda/4$: half wave and quarter wave plate, respectively. L: lens, PD: 25 GHz photodetector.

of the two resonant optical modes is observed by spectrum analyzer and recorded by digital oscilloscope.

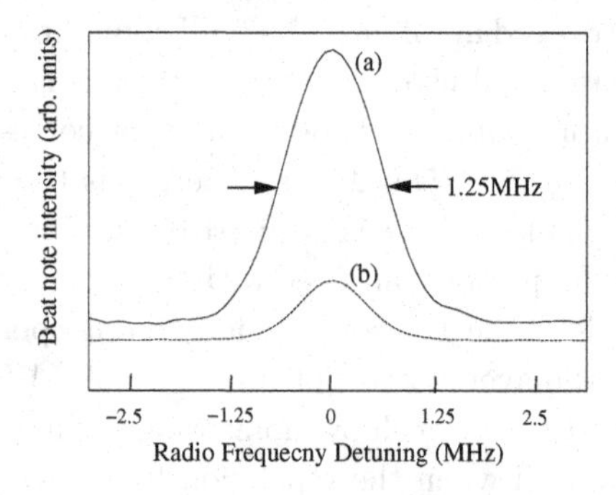

Fig. 8.22 Transmission spectrum of EIT as a function of repetition frequency, the baseline of curve means zero transmission. (a) Laser power 50 μW, (b) laser power 20 μW.

The transmission of mode locked laser is illustrated in Fig. 8.22 with two different laser power. It can be clearly seen that any two optical modes resonant with atomic transition becomes transparent because they drive the atoms into coherent population trapping state which is decoupled from the laser fields. This is a clear manifestation of EIT with transparent window as narrow as 1 MHz, depending on the laser power.

8.5 Nonlinear frequency conversion

Blue-green light can be generated by use of nonlinear crystal to up-conversion of infrared wavelength radiated by high-power semiconductor diode lasers. Second harmonic generation (SHG) is an effective methods to generate the blue-green light, a single infrared diode laser with frequency ν_1 passes through a nonlinear crystal and blue-green light comes out with frequency $2\nu_1$ [Kovlovsky and Lenth (1990); Goldberg *et. al.* (1995); Biaggio *et. al.* (1992)]. These second-order nonlinear effects are relatively weak, yet it is still possible to use them to generate blue-green radiation appropriate for the applications in various fields such as optical data storage, thermal printing, spectroscopy, and display systems. To achieve efficient second harmonic generation, the nonlinear crystal is usually placed into an optical cavity or in an external bow-tie ring cavity [Zimmermann *et. al.* (1995)].

8.5.1 *Second harmonic generation*

It was reported [Brozek *et. al.* (1998)] that 50% conversion efficiency for frequency doubling of a 854 nm GaAlAs diode laser using a potassium notate crystal in an external ring cavity. Frequency stabilization of the diode laser is achieved by direct optical feedback from the doubling cavity or from a grating in Littrow configuration. The blue output power of 7.8 mW shows rams-fluctuations of less than 0.6% in 1 MHz bandwidth. The experimental setup for this system is schematically shown in Fig. 8.23, the radiation of an AR-coated, cw GaAlAs laser diode with a front facet reflectivity of $R < 5 \times 10^{-3}$ is collimated by an AR-coated focusing lens and stabilized by the optical feedback from a holographically fabricated optical grating.

The diode laser system is a typical Littrow configuration. The ring cavity in bow-tie configuration is formed by four mirrors with folding angle as small as possible. The plane mirror M2 is mounted on a fast PZT for scanning the cavity or locking it to the laser frequency, the error signal for locking the cavity to the laser frequency is derived by a polarization spectroscopy scheme [Hansch and Couilaud (1980)]. The crystal is centered around the smaller focus between the two curved mirrors.

As a less expensive alternative, a high power Fabry-Perot diode lase can be used to attain single mode operation and a narrow linewidth in terms of the feedback of transmission of ring cavity into laser after reflecting off a diffraction grating [Sun *et. al.* (2000)]. Experimental setup in the external

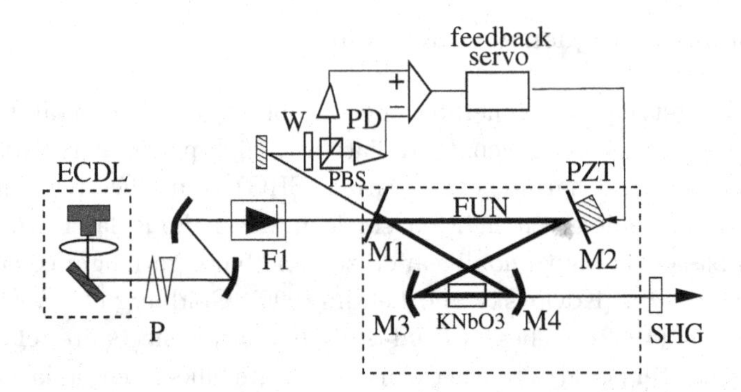

Fig. 8.23 Schematic diagram of the frequency doubling setup. G: grating; P: amyotrophic prism pair; F1: Faraday isolator; PZT: piezo electric transducer; PUBS: polarizing beam splitter; P1,2: photo diode; FUN: fundamental field; SHG: second harmonic generation. Adapted with permission from *Opt. Commun.* **146**, pp. 142. Brozek et al. (1998).

bow-tie ring cavity is shown schematically in Fig. 8.24. The nonlinear crystal is placed in the center between the two spherical mirrors, a larger elliptical beam waist is located between the two plane mirrors, and the diode laser output is focused into this laser waist for mode matching.

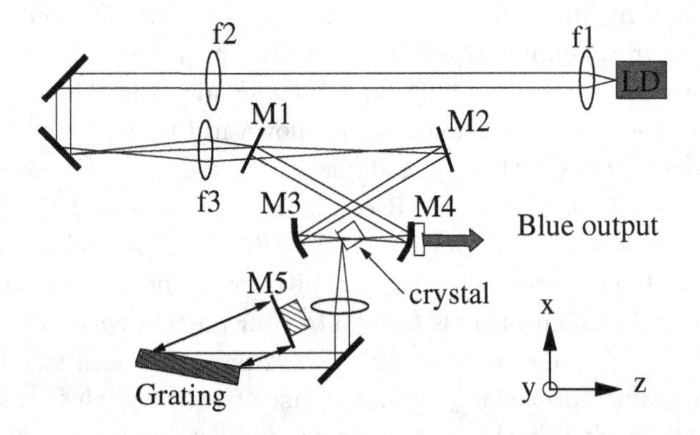

Fig. 8.24 Experimental setup. The bow-tie ring cavity is formed by two plane mirror M1, M2 and two spherical mirrors M3, M4. The reflection off the $KNbO_3$ crystal is reflected back by the grating mirror M5 to provide the feedback for single mode operation and frequency locking. The incident angles on the cavity mirrors are θ, the distance between the two spherical mirror of M3 and M4 is d, and the distance between the two spherical mirrors via plane mirrors (M4-M1-M2-M3) is l. Adapted with permission from *Appl. Phys. Lett.* **76**, 8, pp. 956. Sun et al. (2000).

The InGaAs solitary diode laser is commercial SDL-6530-J1 with threshold 18 mA and maximum output power 200 mW at 250 mA injection current. The spectrum is shown in curve (a) of Fig. 8.25 with an injection current of 240 mA. The single mode spectrum with feedback from the grating is also shown in curve (b). One can see the narrowed spectrum with side mode suppression >40 dB. The output wavelength can be continuously scanned by use of a piezo-driven mirror in one of the cavity mirrors, and changing the cavity length, the feedback phase and the injection current synchronously.

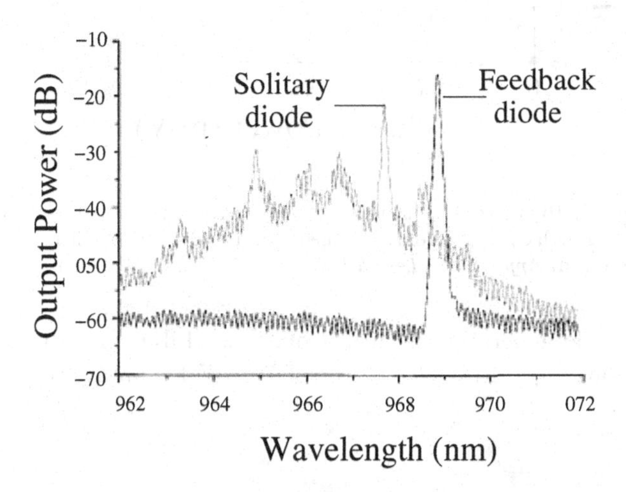

Fig. 8.25 The spectrum of the laser diode without and with feedback. Note that with grating feedback the spectrum narrows and the side mode suppression ratio increases to greater that 40 dB, and the peak wavelength can be tuned by adjusting the angle of the mirror M5. Adapted with permission from *Appl. Phys. Lett.* **76**, 8, pp. 956. Sun et al. (2000).

The blue light output power is plotted as a function of input power of diode lase in Fig. 8.26. The maximum blue output is 6.1 mW at the diode wavelength of 970 nm with nonlinear conversion efficiency $\alpha = P_\omega/P_{2\omega}^2$ of 0.044%/W. The blue light is in a diffraction limited Gaussian TEM_{00} mode and the linewidth is estimated to be less than 15 MHz.

An intracavity frequency-doubling scheme is merged with an actively mode-locked AlGaInP laser diode with an external cavity to provide efficient generation of 335 nm ultraviolet light [Uchiyama and Tsuchiya (1999)]. Intense pulses of 16 ps are successfully generated inside the cavity at subgigahertz repetition frequencies. These pulses are applied to intracavity

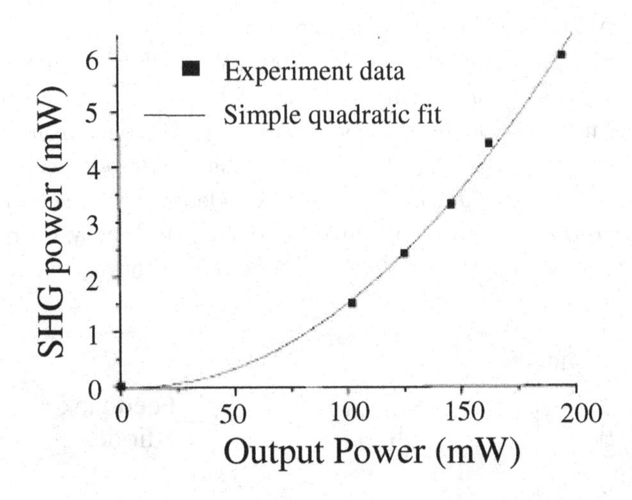

Fig. 8.26 Output SHG power versus output power of the diode laser. The theoretical fit is the quadratic growth expected from SHG theory for low conversion efficiency. Adapted with permission from *Appl. Phys. Lett.* **76**, 8, pp. 957. Sun et al. (2000).

second harmonic generation from a 5 mm long $LiIO_3$ crystal, which resulted in an average ultraviolet power of 79 μW for an average power of 73 mW.

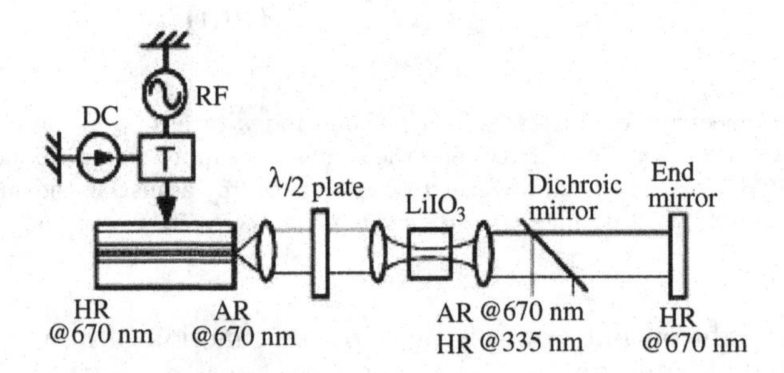

Fig. 8.27 Schematic diagram of the external cavity actively mode-locked AlGaInP LD with intracavity SHG. Adapted with permission from *Opt. Lett.* **24**, 16, pp. 1148-1150. Uchiyama and Tsuchiya (1999).

Figure 8.27 shows schematically the configuration, a 670 nm AlGaInP laser diode with a length of approximately 800 μm. One of its facet is antireflection coated with reflectivity of 5×10^{-4}, the other facet is high-reflection coated. No laser action is obtained without an external cavity.

The external cavity length L is about 69 cm, which gives a round trip frequency f=217 MHz. Both facets of the crystal $LiIO_3$ are AR coated originally for 690 nm and 345 nm light with residual reflectivity of 670 nm and 35 nm light of less than 2%. A half-wave plate is inserted between the laser diode and the crystal. A lens focused the fundamental beam onto the $LiIO_3$ crystal with a spot radius of about 15%. The corresponding second harmonic generation acceptance bandwidth is 0.38 nm. The generated ultraviolet light is taken out of the cavity by a dichroic mirror (DM). The laser is driven by a RF sinusoidal signal with a dc bias current I_b. Its small signal 3 dB bandwidth is limited to 790 MHz. Thus the pulse repetition rate is set near the 434 MHz or the fourth harmonic 870 MHz. Fig. 8.28 shows the dependence of second harmonic power $P_{av}^{2\omega}$ on the input power P_{av}^{ω}. $P_{av}^{2\omega}$=70 μW was obtained for P_{av}^{ω}=73 mW at repetition frequency 870 MHz.

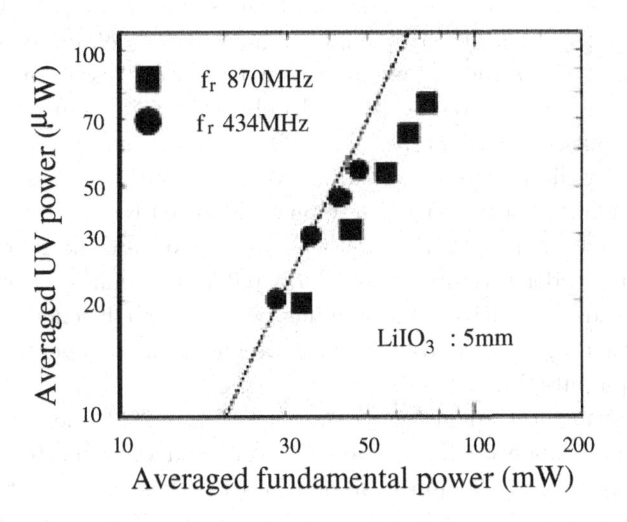

Fig. 8.28 Second-harmonic power $P_{av}^{(2\omega)}$ relative to average fundamental power $P_{av}^{(\omega)}$. Dotted line, square dependence of $P_{av}^{(2\omega)}$ on $P_{av}^{(\omega)}$ for $f_r = 434MHz$. Adapted with permission from *Opt. Lett.* **24**, 16, pp. 1148-1150. Uchiyama and Tsuchiya (1999).

An external cavity 1540 nm diode laser, which is frequency doubled in a 3-cm-long periodically poled $LiNbO_3$ waveguide doubler with 150%/W conversion efficiency, has been reported to generate more than 3μW at 770 nm [Bruner *et. al.* (1998)]. In the recent work of Zimmermann [Zimmermann *et. al.* (2002)], they introduced a new concept for a wavelength-tunable frequency-doubled laser diode with a single control parameter. The concept is based on intracavity frequency doubling in an external resonator geometry with spatial separation of the spectral components. The use of a fan-structured periodically poled LiTaO3 crystal permits tuning of both the fundamental and the second harmonic simultaneously with one aperture. They demonstrated the tunability over more than 10 nm in the blue (480.4 to 490.6 nm) with output powers of the order of 50 nW.

8.5.2 *Frequency quadrupling*

Tunable UV radiation near 215 nm was produced by frequency quadrupling the 860 nm emission of a mode-locked external-cavity compound semiconductor laser containing a tapered GaA1As amplifier. A $KNbO_3$ crystal generated the 430 nm second harmonic, which was doubled by a $\beta - BaB_2O_4$ crystal, producing tunable UV radiation with as much as 15 μW of average power [Goldberg *et. al.* (1995)]. Goldberg group [Kliner *et. al.* (1997)] used similar technique to achieve tunable, narrow-bandwidth less than 200 MHz, ~215 nm radiation by frequency quadrupling the ~ 860nm output of a high-power, pulsed GaAlAs tapered amplifier(TA) seeded by a single-mode external cavity diode laser. Two successive stages of single-pass frequency doubling, the first using noncritically phase-matched $KNbO_3$ and the second using angle-tuned β-barium borate (BBO), convert the fundamental to ~215 nm. Under CW operation, the ~ 0.6 W maximum output power of the TA leads to relatively low nonlinear conversion efficiency. The conversion efficiency is increased by 2 orders of magnitude by operation of the amplifier in a pulsed mode.

The experimental apparatus is shown in Fig. 8.29. The single-mode external cavity diode laser of power >20 mW and wavelength tunability between 842 nm and 868nm is seeded into the TA centered at 852 nm with gain >30 nm. The TA is driven with either a CW or a pulsed current source. The TA output is collimated by spherical and cylindrical lenses, then focused into a 1.24-cm-long, temperature controlled, α-cut $KNbO_3$ crystal with AR coated for 860 nm on input face and 430 nm on the output face. The generated blue beam is refocused with an f=5 cm lens into a 4-mm-long, MgF_2-coated BBO crystal, which is cut at 70 ° and rotated the 70.6 ° propagation angle required for type I phase matching. The UV beam emerging from the BBO crystal is separated from the IR and blue beam by a CaF_2 prism, dichroic beam splitters, a $NiSO_4 \cdot H_2O$ filter, then is detected with a Si photodiode.

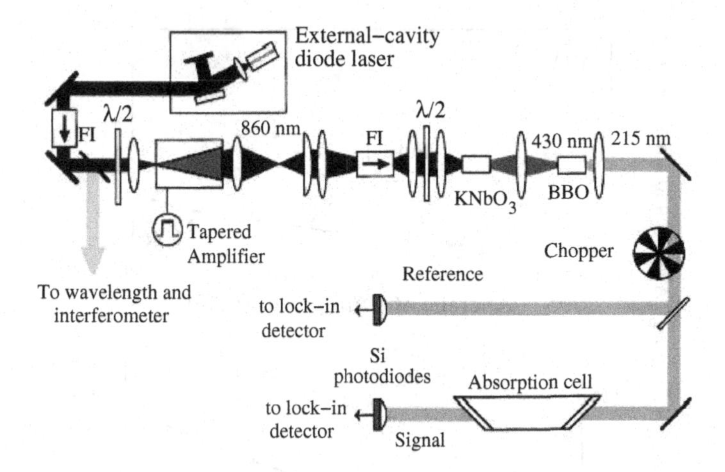

Fig. 8.29 Schematic of the experimental apparatus, showing the UV laser system and the setup used to record absorption spectra. FIs: Faraday isolators; $\lambda/2$, half-wave plate. Reprinted with permission from *Opt. Lett.* **22**, 18, pp. 1418-1420. Kliner et al. (1997).

Figure 8.30 shows the average power of blue and UV as function of power of infrared for both cw and pulsed operation. The laser system has a tuning range of 842$\sim$868 nm corresponding to a UV tuning range of 210$\sim$217 nm with linewidth <200 MHz, and well suited for spectroscopic detection methods discussed later in this chapter.

Recently, Sayama and Ohtsu [Sayama and Ohtsu (1998)] have developed a tunable ultraviolet cw light source by frequency up-conversion of laser diodes generated by a BBO crystal placed in an external resonant enhancement cavity in terms of sum-frequency mixing between the 778 nm output of a diode laser and the second harmonic of the 857.2 nm output of laser diode which is generated in a $KNbO_3$ crystal placed in an external resonant enhancement cavity. The ultraviolet power of 35.7 nW is obtained at 276.4 nm wavelength with input fundamental power of 50 mW at 778 nm at 857.2 nm. The tuning range of the ultraviolet generation is from 275.5 nm to 276.4 nm. This wavelength is appropriate for laser induced fluorescence detection of silicon.

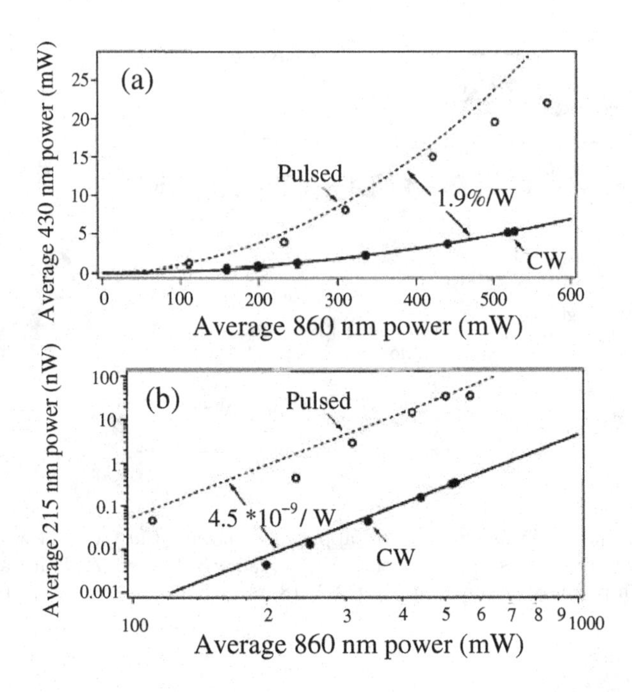

Fig. 8.30 Average blue (a) and UV (b) power versus average IR power delivered to the $KNbO_3$ crystal for cw and pulsed operation of the TA. (a) The curve corresponds to an IR → blue conversion efficiency of α_{blue}=1.9%/W. The blue power was measured immediately after the $KNbO_3$ crystal. (b) The lines corresponds to an IR → UV conversion efficiency of $\alpha_{UV} = 4.5 \times 10^{-9}/W$. The UV power has been corrected for Fresnel losses on the optics and for absorption by the $NiSO_4 \cdot H_2O$ filter. Reprinted with permission from *Opt. Lett.* **22**, 18, pp. 1418-1420. Kliner et al. (1997).

8.6 Optical telecommunication

8.6.1 *Coherent system and DWDM*

ECDLs have been shown to offer the single-mode operation, narrow spectral linewidth and wide tunability, which are required by highly coherent phase-shift keying (PSK) and frequency-shift keying (FSK) optical communication systems [Creaner *et. al.* (1988)]. Today this is one of the most important areas for the application of these lasers. The development of more and more sophisticated diode laser systems has made new concepts in optical communication practical [Coldren (2003)].

Mellis et al.[Mellis *et. al.* (1988)] have developed miniature, packaged, grating-external-cavity lasers for use in PSK and FSK coherent transmission systems at 1.5 μm. The lasing wavelength is mechanically adjustable over the range 1550-1560 nm and electrically tunable over a range of 0.8 nm and spectral linewidth <100 kHz, With appropriate control of the grating piezoelectrical transducer, a continuous (mode hop-free) tuning range of 50 GHz has been attained.

External cavity diode lasers exhibit superior performance and are well suited for deployment in dynamic dense wavelength division multiplexing (DWDM) networks. In particular, the ECDL design has the best performance when it comes to wide tunability, high power, narrow linewidth and low noise. The applications of tunable lasers in DWDM networks are numerous, stretching from hot sparing and inventory management to dynamic add-drop multiplexes and transponders for real-time provisioning and protection. With the development of the techniques of widely tunable external-cavity diode lasers and diffraction-grating filters based on silicon micro-electro-mechanical-system (MEMS) actuators. This technique has been used in dense wavelength division multiplexing (DWDM) telecommunications [Berger and Anthon (2003)].

8.6.2 *Testing and measurement*

Development of the DWDM technology leads to the need for accurate wavelength measurement and standards to make sure compatibility of equipment from different manufacturers. For DWDM systems, frequency spacing equal to an integral multiple of 100 GHz is allowed. The reference sources can be used to calibrate the accuracy of important DWDM diagnostic tools such as wavelength meters and high-resolution spectrometers, which are then used to check the passive components such as the fiber grating and received filters. These diagnostic tools also can measure the wavelength of the stabilized distributed feedback diodes and fiber grating controlled laser-diode sources used in DWDM [Lang (1998)]. The most accurate measurement of these sources is made by direct measurement against the DWDM reference source. The laser being measured can then be traced back to the cesium clock with megahertz precision.

Fig. 8.31 schematically shows heterodyne measurement, the output of the DWDM reference source and the test laser are coupled into a high-speed InGaAs photodetector, The resulting beat frequency is measured with a high-accuracy spectrum analyzer or frequency counter. One method of providing an absolute yet compact wavelength standard is to lock the output of a diode laser to an atomic or molecular spectral line. However,

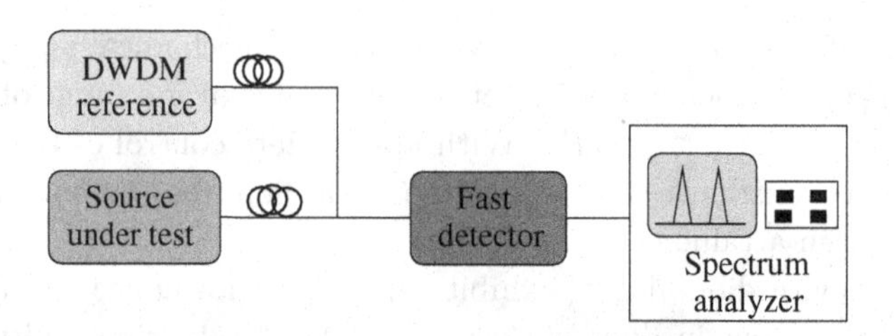

Fig. 8.31 Wavelength-division-multiplexing source is calibrated by measuring a heterodyne beat frequency using a high-speed InGaAs detector. Adapted with permission from *Laser Focus World*, June. Lang (1998).

there are almost no spectral standards in the 1550 nm communication window spectral region, the rapid growth of DWDM technology has changed this situation.

Tremendous advances have been made in calibrating the spectral of both acetylene (C_2H_2) and hydrogen cyanide (HCN) because these molecules have a combination of vibrational overture absorption lines that span the entire 1550 nm erbium gain region as shown in Fig. 8.32. Unlike the single atomic lines of neon and krypton, these spectra have a comb of spectral lines throughout the spectral region of communication interest. Importantly, their absorption can be detected in simple, compact cells. So acetylene and hydrogen cyanide absorption lines have been chosen as the basis for a series of rugged reference sources for DWDM work. (see Fig. 8.33) [Lang (1998)]. The emitting element is a laser diode in a Littman-Metcalf external cavity, the collimator output of this cavity is coupled into a fiber coupler. A beam splitter, positioned before this coupler, splits off 5% of laser , which allows a normalized measurement of the absorption in a cell containing C_2H_2 or HCN, by a pair of InGaAs photo detectors. There are many rotational lines in the C_2H_2 and HCN spectra to which the DWDM reference source can be stabilized. In practice, the end user selects from the list of absolute wavelengths.

8.7 Other applications

Recently optical gas detector based on semiconductor diode lasers have found applications for a variety of environmental, medical and industrial process. Their characteristics like remote sensing, gas selectivity, high detection speed and low cost make them more appealing than conventional

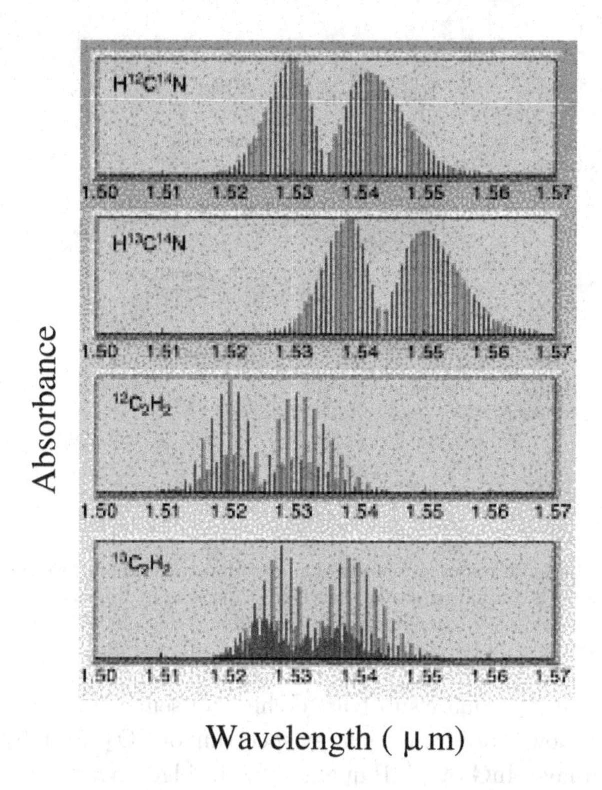

Fig. 8.32 Vibrational overtune bands of hydrogen cyanide (HCN) and C_2H_2 together span the erbium gain region. Adapted with permission from *Laser Focus World*, June. Lang (1998).

electrochemical and semiconductor point sensors. In this section, we introduce the gas monitoring sensor based on external cavity diode laser and briefly mention the LIDAR system.

8.7.1 *Gas monitoring sensor*

Near-IR room temperature ECDLs have been used for *in situ* absolute measurements of species concentrations and gas temperature in a variety of combustion system [Baer *et. al.* (1994)]. Development conducted by Focused Research has extended the current available wavelength range of 0.4-1.6 μm ECDLs to 2 μm external-cavity diode laser. This wavelength region is especially useful for monitoring the molecular species such as CO_2,H_2O, N_2O, and NH_3 for combustion diagnostics and environmental monitoring

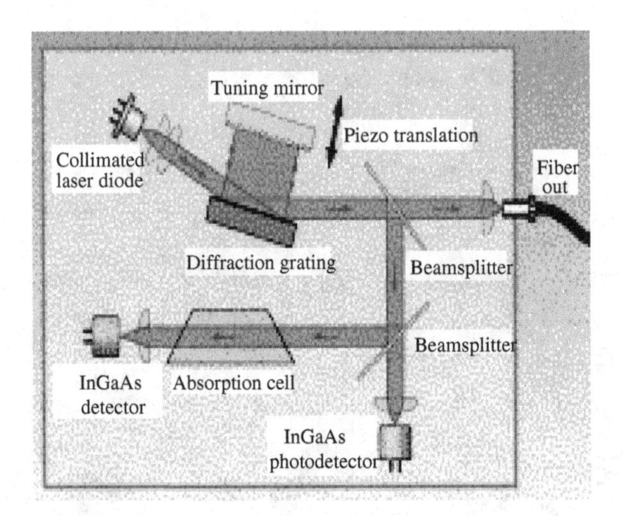

Fig. 8.33 DWDM reference source consists of an external-cavity diode laser locked to an integral absorption cell containing either acetylene (C_2H_2) or hydrogen cyanide (HCN). Adapted with permission from *Laser Focus World*, June. Lang (1998).

and HBr for *in-situ* gas-phase substrate etching for semiconductor industry.

Figure 8.34 shows data from a survey spectrum of CO_2 taken by ECDLs with a strained-layer InGaAs/InP quantum-well ridge-wave-guide semiconductor laser and in the configuration of Littman-Metcalf. A number of UV

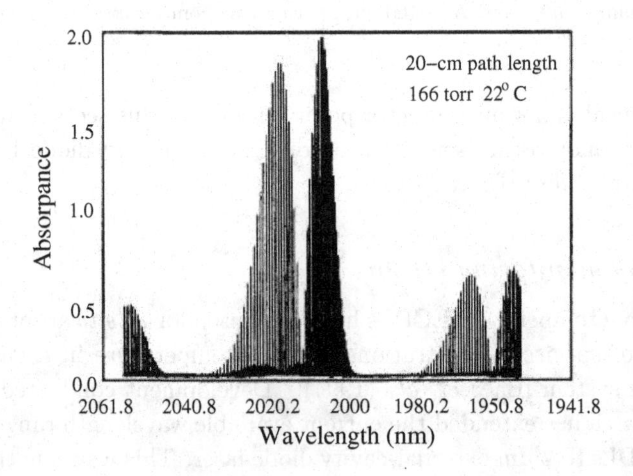

Fig. 8.34 Survey spectrum of CO_2 taken with a 2 μ m ECDL.

absorption measurements for various species have been reported [Oh (1995); Ray *et. al.* (2001)]. Koplow et al. used the technique as shown in Fig. 8.29 to UV light for NO absorption measurements [Koplow *et. al.* (1998)]. Peterson and Oh used second harmonic generation of the output of a 860 nm diode laser for absorption and laser induced fluorescence measurement for CH radical [Peterson and Oh (1999)]. Alnis et al. used sum-frequency mixing of the output of a newly developed blue diode laser at 404 nm and a 688 nm diode laser to generate radiation at 254 nm for spectroscopy of mercury atom [Alnis *et. al.* (2000)]. A monitor for Al vapor density based on atomic absorption using frequency-doubled external-cavity-diode laser source at 394 nm has been demonstrated in both evaporation and sputtering processes [Wang *et. al.* (1994)]. Closed loop operation was achieved for electron beam evaporated aluminum in a vacuum chamber using the atomic absorption signal for feedback.

8.7.2 *LIDAR*

LIDAR is an acronym for "LIght Detection And Ranging", it is similar to the well-known RADAR. A LIDAR sends out short laser pulses into the atmosphere. All along its path the light is scattered by air molecules (Rayleigh scattering) and by particles. A small fraction of the light is backscattered to the LIDAR system and is received by a telescope and a sensitive detector integrated in the system. The received signal is acquired as a function of time. Given the constant value of the velocity of light and the time that the emitted and backscattered light needs until its detection, it is possible to get information about the spatial distribution of molecules and aerosols along the beam path.

The group of Texas A & M University has been working on a project to rapidly an accurately monitor upper-ocean vertical sound velocity and temperature profiles using Brillouin scattering. When sound waves pass through a refractive medium such as sea water, they create periodic fluctuations in density. When light is scattered by a moving sound wave in this medium, the periodic density fluctuation act as a moving diffraction grating. This diffraction is referred to as Brillouin scattering. Conservation of momentum dictates that the scattered light is Doppler-shifted, producing a blue and red shift component relative to the incident beam. This spectral shift is directly determined by the sound velocity, which in turn can be used to infer the local temperature. The narrow-linewidth high-energy pulses will be produced by seeding a Nd:YAG laser with an ultrastable external-cavity diode laser and doubling the output to produce 532 nm, which transmits well through the see water. The novel, yet simple detection method to measure Brillouin shifts uses edge absorption of an iodine absorption line. The integrated iodine fluorescence intensity is a direct measure of the magnitude of the Brillouin shift as illustrated in Fig. 8.35.

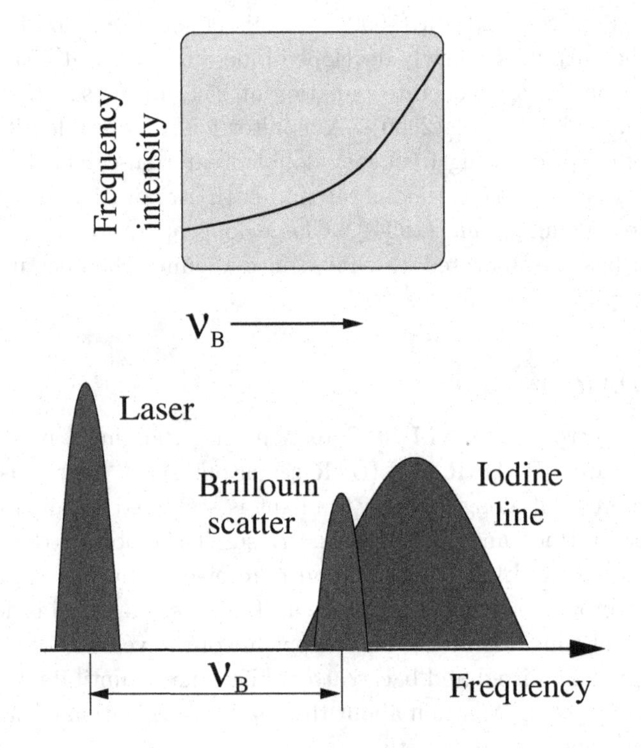

Fig. 8.35 ECDL is used to measure the edge absorption, in which the integrated iodine fluorescence (inset) is determined by the magnitude of the Brillouin shifted. Adapted with permission from *Laser Focus World*, June. Lang (1998).

Chapter 9

Conclusions

It has been shown through the evidence presented in this book that tunable external cavity diode lasers will have a considerable impact on development of the next generation of optical networks and scientific research in atomic and molecular physics and engineering. This promises to open up a new avenue to acquire coherent radiation sources with wide continuous tuning range, high output power, excellent wavelength accuracy and reliability, narrow linewidth and good single mode suppression. The main results in this book can be summarized as follows:

- The basic concepts of semiconductor diode lasers with a variety of structures from bulk to quantum dots lasers have been introduced to describe the operation principles of tunable diode lasers.
- Monolithic tunable semiconductor lasers with the latest configurations have been examined with emphasis on their tuning range and spectral purity, and their performance with respect to the potential applications has been discussed. Relevant monolithic tunable lasers are based on DFB and DBR-type lasers and their derivatives.
- Elements, systems, and applications of tunable external cavity diode lasers are our main concern. A broad range of tunable external cavity diode lasers has been explored from their basic characteristics to a complicated system configuration. Wide tunability with various approaches has been introduced in mechanical and electronic terms. Applications from coherent fiber communications to manipulation of atoms are covered for industrial and scientific significance.
- Recent developments in optical technology make it necessary to use frequency stabilized diode lasers, especially for the technology of dense wavelength division multiplex and quantum optical engineering. we have developed different methods to stabilize external cavity diode

lasers to meet these marginal requirements.

New applications are arising that require tunable external cavity diode lasers capable of compact, reliable and wide tunability in the gain bandwidth, and that produce fast linear frequency switching, which will be key enablers for future wavelength-agile systems as well as for atomic physics, remote sensing, and environmental monitoring. The tendency of modern technology is for the ultimate user to want a tunable external cavity diode like a module that includes a complete control circuit designed in such a way that the user need only to hit the button and get the desired wavelength and output power. There are a couple of companies designing this control circuit and algorithms for the next generation devices.

Bibliography

Affolderbach, C., Stahler, M., Knappe, S. and Wynands, R. (2002). An all-optical, high sensitivity magnetic gradiometer, *Appl. Phys. B* **75**, pp. 605-612.

Agrawal, G. P. (1992). *Fiber optic communication systems*, Academic Press, Boston.

Agrawal, G. P. and Dutta, N. K. (1993). *Semiconductor lasers*, New York, Van Nostrand Reinhold.

Akerman, D., Elisseev, P. G., Kaiper, A., Manko, M. A. and Roab, Z. (1971). *Sov. J. Quantum Electron.* **1**, pp. 60.

Al-Chalabi, S. A., Mellis, J., Hollier, M., Cameron, K. H., Wyatt, R., Regnault, J. E., Delvin, W. J. and Brain, M. C. (1990). Temperature and mechanical vibration characteristics of a miniature long external cavity semiconductor laser, *Electron. Lett.* **26**, 15, pp. 1159-1160.

Aleksandrov, E. B., Balabas, M. V., Pasgalev, A. S., Vershovskii, A. K. and Yakobson, N. N. (1996). *Laser Physics*, **6**, pp. 244.

Alferness, R. C., and Buhl, L. L. (1982). Tunable electron-optic waveguide TE↔TM converter/wavelength filter, *Appl. Phys. Lett.* **40**, pp. 861-862.

Alferness, R. C., Koren, U., Buhl, L. L., Miller, B. I., Young, M. G., Koch, T. L., Raybon, G. and Burrus, C. A. (1992). Broad tunable InGaAsP/InP laser based on a vertical coupler filter with 57nm tuning range, *Appl. Phys. Lett.* **60**, 26, pp. 3209-3211.

Alferov, Zh. I., Andreev, V. M., Portnoi, E. L. and Trukan, M. K. (1970). AlAs-GaAs heterojunction injection lasers with a low room-temperature threshold, *Fiz. Tekh. Polupovodn* **3**, 9, pp. 1328-1332.

Allard, F., Maksimovic, I., Abgrall, M. and Laurent, Ph. (2004). Automatic system to control the operation of an extended cavity diode laser, *Rev. Sci. Instrum.* **75**, 1, pp. 54-58.

Alnis, J., Gustafsson, U., Somesfalean, G. and Svanberg, S. (2000). Sum-frequency generation with a blue diode laser for mercury spectroscopy at 254 nm, *Appl. Phys. Lett.* **76**, pp. 1234-1236.

Alzetta, G., Gozzini, A., Moi, L. and Orriols, G. (1976) *Nuovo Cimento B* **36**, pp. 5.

Amann, M. C. and Buss, J. (1998) *Tunable laser diodes*, London, Artech House.

Andalkar, A., Lamoreaux, S. K. and Warrington, R. B. (2000). Improved external cavity design for cesium $D1(894\ nm)$ diode laser, *Rev. Sci. Instrum.* **71**, 11, pp. 4029-4031.

Andrews, J. R. (1991). Electronically tunable single-mode external-cavity diode laser, *Opt. Lett.* **16**, 10, pp. 732-734.

Arditi, M. and Picque, J. L. (1975). Application of the light shift effect to laser frequency stabilization with reference to a microwave frequency standard, *Opt. Commun.* **15**, pp. 317-322.

Arie, A., Schiller, S. S., Gustafson, E. K. and Byer, R. L. (1992). Absolute frequency stabilization of diode-laser-pumped Nd:YAG lasers to hyperfine transitions in molecular iodine, *Opt. Lett.* **17**, 17, pp. 1204-1206.

Arimondo, E. (1996). In *Prog. Opt.* **35**, Edited by E. Wolf, Elsevier Science, Amsterdam, pp. 257.

Arnold, A., Wilson, J. S. and Boshier, M. G. (1998). A simple extended-cavity diode laser, *Rev. Sci. Instrum.* **69**, 3, pp. 1236-1239.

Atutov, S. N., Mariotti, E., Meucci, M., Bicchi, P., Marinelli, C. and Moi, L. (1994). A 670 nm external-cavity single mode diode laser continuously tunable over 18 GHz range, *Opt. Commun.* **107**, pp. 83-87.

Baer, D. S., Hanson, R. K., Newfield, M. E. and Gopaul, N. K. (1994). Multiplexed diode-laser sensor system for simultaneous H_2O, O_2 and temperature measuremnets, *Opt. Lett.* **19**, 22, pp. 1900-1902.

Bagley, M., Wyatt, R., Elton, D. J., Wickes, H. J., Spurdens, P. C., Seltzer, C. P., Cooper, D. M. and Devlin, W. J. (1990). 242 nm continuous tuning from a GRIN-SC-MQW-BH InGaAsP laser in an extended cavity, *Eletron. Lett.* **26**, 4, pp. 267-269.

Bai, Y. S., Babbitt, W. R. and Mossberg, T. W. (1986). Cohernet transient optical pulse-shape storage/recall using frequency swept excitation pulses, *Opt. Lett.* **11**, pp. 724-726.

Baluschev, S., Friedman, N., Khaykovich, L., Carasso, D., Johns, B. and Davidson, N. Tunable and frequency-stabilized diode laser with a Doppler-free two-photon Zeeman lock, *Appl. Opt.* **39**, 27, pp. 4970-4974.

Basov, N. G., Krokhin, O. N. and Popov, Yu. M. (1961). Production of negative temperature states in P-N junctions of degenerate semiconductors, *JETP.* **40**, pp.1320.

Bayram, S. B. and Chupp, T. E. (2002). Operation of a single mode external-cavity laser diodes array near 780 nm, *Appl. Phys. Lett.* **73**, 12, pp. 4169-4171.

Berger, J. D. and Anthon, D. (2003). Tunable MEMS devices for optical networks, *Optics & Photonics News*, March, pp. 43-49.

Bernard, M. G. A. and Duraffoug, D. (1961). Laser condition in semiconductors, *Phys. Status Solidi*, **1**, pp. 699.

Besnard, P., Meziane, B. and Stephan, G. M. (1993). Feedback phenomena in a semiconductor lasers induced by distant reflectors, *IEEE J. Quantum Electron.* **29**, 5, pp. 1271-1284.

Bhatia, P. S., Wlech, G. R. and Scully, M. O. (2001). A single-mode semiconductor diode laser operating in the strong optical feedback regime and tunable within the D_1 line of the Cs atom, *Opt. Commun.* **189**, pp. 321-336.

Biaggio, B., Kerkoc, P., Wu, L. S., Gunter, P. and Zysset, B. (1992). Refractive indices of orthorhomic $KNbO_3$ II, Phase-matching configurations for nonlinear-optical interactions, *J. Opt. Soc. Am. B* **9**, 4, pp. 507.

Bird, D. M., Armitage, J. R., Kashyap, R., Fatah, R. M. A. and Cameron, K. H. (1991). Narrow line semiconductor laser using fiber grating, *Eletron. Lett.* **27**, 13, pp. 1115-1116.

Boggs, B., Greiner, G., Wang, T., Lin, H. and Mossberg, T. M. (1998). Simple highly-cohernece rapid tunable external-cavity diode laser, *Opt. Lett.* **23**, 24, pp. 1906-1908.

Boshier, M. G., Berkeland, D., Hinds, E. A. and Sandoghdar, V. (1991). External cavity frequency-stabilization of visible and infrared semiconductor lasers for high resolution spectroscopy, *Opt. Commun.* **85**, pp. 355-359.

Bottger, T., Pryde, G. J. and Cone, R. L. (2003). Programmable laser frequency stabilization at 1523 nm by use of persistent spectral hole burning, *Opt. Lett.* **28**, 3, pp. 200-202.

Bowers, J. E., Morton, P. A., Mar, A. and Corzine, S. W. (1989). Actively mode-locked semiconductor lasers, *IEEE J. Quantum Electron.* **25**, 6, pp. 1426-1439.

Boyd, G. D. and Heismann, F. (1989). Tunable acoustooptic reflection filter in $LiNbO_3$ without a Doppler shift, *J. Lightwave Technol.* **7**, 4, pp. 625-631.

Bradley, C. C., Chen, J. and Hulet, R. G. (1990). Instrumentation for the stable operation of laser diodes, *Rev. Sci. Instrum.* **61**, 8, pp. 2097-2101.

Breede, M., Hoffmann, S., Zimmermann, J., Struckmeier, J., Hofmann, M. et al. (2002). Fourier-transfer external cavity lasers, *Opt. Commun.* **207**, pp. 261-271.

Brozek, O. S., Quetschke, V., Wicht, A. and Danzmann, K. (1998). Highly efficienct cw frequency doubling of 854 *nm* GaAlAs diode lasers in an external ring cavity, *Opt. Commun.* **146**, pp. 141-146.

Bruner, A., Arie, A., Arbore, M. A. and Fejer, M. (1998). Frequency stabilization of a diode laser at 1540 nm by locking to sub-Doppler lines of potassium at 770 nm, *Appl. Opt.* **37**, 6, pp. 1049-1052.

Budker, D., Kimball, D. F., Rochester, S. M., Yashchuk, V. V. and Zolotorev, M. (2000). Sensitive magnetometry based on nonlinear magneto-opticcal rotation, *Phys. Rev. A* **62**, pp. 043403.

Budker, D., Gawlik, W., Kimball, D. F., Rochester, S. M., Yashchuk, V. V. and Weis, A. (2002). Resonant nonlinear magneto-optical effects in atoms, *Rev. Mod. Phys.* **74**, 4, pp. 1153-1201.

Buus, J. (1991). *Single frequency semiconductor lasers*, Bellingham, WA, SPIE.

Bykovski, Yu, A. et al. (1970). Use of a Faby-Perot resonator for stabilization of the frequency of an injection laser, *Sov. Phys. Semiconductors* **4**, pp. 580.

Cafferty M. S. and Thompson, E. D. (1989). Stable current supply with protection circuits for a lead-salt laser diode, *Rev. Sci. Instrum.* **60**, pp. 2896-2901.

Camparo, J. C. (1985). *Contemp. Phys. Methods.* **26**, pp. 443.

Carroll, J., Whiteaway, J. and Plumb, D. (1998). *Distributed feedback semiconductor lasers*, SPIE Press **52**.

Cassettari, D., Arimondo, E. and Verkerk, P. (1998). External-cavity broad-area laser diode operating on the D_1 line of cesium, *Opt. Lett.* **23**, 14, pp. 1135-1137.

Cavelier, M., Stelmakh, N., Xie, J. M., Chusseau, L., Lourtioz, J. M., Kazmierski, C. and Bouadma, N. (1992). Picosecond (≤ 2.5 ps) wavelength-tunable ($\sim$ 20 nm) semiconductor laser pulses with repetition rates up to 12 GHz, *Electron. Lett.* **28**, 3, pp. 224-226.

Chang-Hasnain, C. J. (2000). Tunable VCSEL, *IEEE J. Select. Topics on Quantum Electron.* **6**, 6, pp. 978-987.

Chawki, M. J., Valiente, I., Auffret, R. and Tholey, V. (1993). All fiber 1.5 μm widely tunable single frequency and narrow linewidth semiconductor ring laser with fiber Fabry Perot filter, *Electron. Lett.* **29**, 23, pp. 2034-2035.

Chen, J., Sibbett, W. and Vukusic, J. I. (1984). Tunable mode-locked semiconductor lasers incorporating Brewster-angled diode, *Opt. Commun.* **48**, 6, pp. 427-431.

Chen, Y. T. (1989). *Appl. Opt.* **28**, pp. 2017.

Cheron, B., Gilles, H., Havel, J., Moreau, O. and Sorel, H. (1994). *J. Phys. III*, **4**, pp. 401.

Chuang, Z. M. and Coldren, L. A. (1993). Design of widely tunable semiconductor lasers using grating-assisted codirectional-couplers filters, *IEEE J. Quantum Electron.* **29**, 4, pp. 1071-1080.

Clifford, M. A., Lancaster, G. P. T., Conroy, R. S. and Dholakia, K. (2000). Stabilization of an 852 nm extended-cavity diode laser using the Zeeman effect, *J. Mod. Opt.* **47**, 11, pp. 1933-1940.

Coldren, L. A. and Koch, T. L. (1984). Analysis and design of coupled-cavity lasers–Part I: Threshold gain analysis and design guidelines, *IEEE J. Quantum Electron.* **20**, 6, pp. 659-670.

Coldren, L. A. and Corzine, S. W. (1987). Continuously-tunable single-frequency semiconductoe lasers, *IEEE J. Quantum Electron.* **23**, 6, pp. 903-908.

Coldren, L. A. and Corzine, S. W. (1995). *Diode lasers and photonic integrated circuits*, New York: Wiley.

Coldren, L. A. (2000). Monolithic tunable didoe lasers, *IEEE J. Selected Topics on Quantum Electron.* **6**, 6, pp. 988-999.

Coldren, L. A. (2003). *Tunable semiconductor lasers*, OFC tutorials.

Conroy, R. S., Carleton, A. and Dholakia, K. (1999). A compact high-performace extended-cavity diode laser at 635 nm, *J. Mod. Opt.* **46**, 12, pp. 1787-1791.

Conroy, R. S., Hewett, J. J., Lancaster, G. P. T., Sibbert, W., Allen, J. W. and Dholakia, K. (2000). Characterisation of an extended cavity violet diode laser, *Opt. Commun.* **175**, pp. 185-188.

Coquin, G. A. and Cheung, K. W. (1988). Electronically tunable external-cavity semiconductor laser, *Electron. Lett.* **24**, 10, pp. 599-600.

Coquin, G. A., Cheung, K. W. and Choy, M. M. (1989). Single- and multi-wavelength operation of acoustooptically tuned semiconductor laser at 1.3 μm, *IEEE J. Quantum Electron.* **25**, 6, pp. 1575-1579.

Corwin, K. L., Lu, Z. T., Hand, C. F., Epstein, R. J. and Wieman, C. E. (1998). Frequency-stabilized diode laser with the zeeman shift in an atomic vapor,

Appl. Opt. **37**, 15, pp. 3295-3298.
Corzine, S. W., Bowers, J. E., Przybylek, G., Koren, U., Miller, B. I. and Soccolich, C. E. (1988). Actively mode-locked GaInAsP laser with subpicosecond output, *Appl. Phys. Lett.* **52**, 5, pp. 348-350.
Creaner, M. J., Steele, R. C., Marshall, I., Walker, G. R., Walker, N. G., Mellis, J., Al Chalabi, S., Sturgess, I., Rutherford, M., Davidson, J. and Brain, M. (1988). Field demonstration of 565 Mbit/s DPSK coherent transmission system ove 176 km of installed fiber, *Electron. Lett.* **24**, 22, pp. 1354-1356.
Crowe, J. W. and Craig, Jr., R. M. (1964). *Appl. Phys. Lett.* **5**, pp. 72-74.
Dahmani, B., Hollberg, L. and Drullinger R. (1987). Frequency stabilization of semiconductor lasers by resonant optical feedback, *Opt. Lett.* **12**, 11, pp. 876-878.
Day, T., Brownell, M. and Wu, I. F. (1995). *Proc. SPIE* **2378**, pp. 35.
Debregeas-Silard, H., Vuong, A., Delmore, F., David, J., Allard, V., Bodere, A., LeGouezigou, O., Gaborit, F., Rotte, J., Goix, M., Voiriot, V. and Jacquet, J. (2002). DBR module with 20 mW coupled output power, over 16 nm (40×50 GHz) spaced channels, *IEEE Photo. Tech. Lett.* **13**, 1, pp. 4-6.
de Labachelerie, M. and Cerez, P. (1985). An 850 nm semiconductor laser tuning over a 300 A range, *Opt. commun.* **55**, 3, pp. 174-178.
de Labachelerie, M. and Passedat, G. (1993). Mod-hop suppression of Littrow grating-tuned lasers, *Appl. Opt.* **32**, 3, pp. 269-274.
Delfyett, P. J., Florez, L. T., Stoffel, N., Gmitter, T., Aandreadakis, N. C., Silberg, Y., Heritage, J. P. and Alphonse, G. A. (1992). High-power ultrafast laser diodes, *IEEE J. Quantum Electron.* **28**, 10, pp. 2203-2219.
Dente, G. C., Durkin, P. S., Wilson, K. A. and Moeller, C. E. (1988). Chaos in the coherence collapse of semiconductor lasers, *IEEE J. Quantum Electron.* **24**, 12, pp. 2441-2447.
Dingle, R., Wiegmann, W. and Henry, C. H. (1974). Quantum states of confined carriers in very thin $Al_xGa_{1-x}As - GaAs - Al_xGa_{1-x}As$ heterostructures, *Phys. Rev. Lett.* **33**, 14, pp. 827-830.
Dinneen T. P., Wallace, C. D. and Gould, P. L. (1992). Narrow linewidth, highly stable, tunable diode laser system, *Opt. Commun.* **92**, pp. 277-282.
Dratler, Jr., J. (1974). A proportional thermostat with 10 microdegree stability, *Rev. Sci. Instrum.* **45**, pp. 1435-1437.
Drever, R. W. P., Hall, J. L., Kowalski, F. V., Hough, J., Ford, G. M., Munley, A. J. and Ward, H. (1983). Laser phase and frequency stabilization using and optical resonator, *Appl. Phys. B.* **31**, pp. 97.
Duarte, F. J. (1996). *Tuanble lasers handbook,* Academic Press,
Dupuis, R. D., Dapkus, P. D., Holonyak, Jr., N. and Kolbas, R. M. (1979). Continuous room-temperature multiple-quantum-well $Al_xGa_{1-x}As-GaAs$ injection lasers grown by metal organic vapor depositon, *Appl. Phys. Lett.* **35**, 7, pp. 487-489.
Dutta, N. K., Piccirilli, A. B., Cella, T. and Brown, R. L. (1986). Electronically tunable distributed feedback lasers, *Appl. Phys. Lett.* **48**, 22, pp. 1501-1503.
Dutta, N. K., Cella, T., Piccirilli, A. B. and Brown, R. L. (1986). Integrated external cavity laser, *Appl. Phys. Lett.* **49**, 19, pp. 1227-1229.

Eliseev, P. G., Ismailov, I. and Manko, M. R. (1969). *JETP Lett.* **9**, pp. 362.

Esman, R. D. and Rode, D. L. (1983). 100 *mu*K temperature controller, *Rev. Sci. Instrum.* **54**, 10, pp. 1368-1370.

Favre, F., Le Guen, D., Simon, J. C. and Landousies, B. (1986). External-cavity semiconductor laser with 15 nm continuous tuning range, *Electron. Lett.* **22**, pp. 795-796.

Favre, F. and Le Guen, D. (1991). 82 nm of continuous tunability for an external cavity semiconductor laser, *Electron. Lett.* **27**, 2, pp. 183-184.

Figger, H., Meschede, D. and Zimmermann, C. (2002). *Laser physics at the limits*, Springer-Verlag.

Fish, G. A. (2001). Monolithic widely tunable DBR lasers, in *Proc. OFC*, Anaheim, CA, paper TuB1.

Fischer, I., van Tartwijk, G. H. M., Levine, A. M., Elsasser, W., Gobel, E. and Lenstra, D. (1996). Fast pulsing and chaotic itinerancy with a drift in the coherence collapse of semiconductor laser, *Phys. Rev. Lett.* **76**, 2, pp. 220-223.

Fitzgerald, R. (2003). New atomic magnetometer achieves subfemtotesla sensitivy, *Physics Today*, July **56**, 7, pp. 21-23.

Fleischhauer, M. and Scully, M. O. (1994). Quantum sensitivity limits of an optical magnetometer based on atomic phase coherence, *Phys. Rev. A* **49**, 3, pp. 1973.

Fleming, M. W. and Mooradian, A. (1981). Spectral characteristics of external-cavity controlled semiconductor lasers, *IEEE J. Quantum Electron.* **QE-17**, 1, pp. 44-59.

Fox, R. W., Hollberg, L. and Zibrov, A. S. (1997). Semiconductor diode lasers, in *Experimental Methods in the Physics Sciences* **29C**, Atomic, Molecular and Optical Physics, Academic Press, pp. 77-102.

Gee, S. and Bowers, J. E. (2001). Ultraviolet picosecond optical pulse generation from a mode-locked InGaN laser diode, *Appl. Phys. Lett.* **79**, 13, pp. 1951-1952.

Gilbert, S. L., Etzel, S. M. and Swann, W. C. (2002). Wavelength accuracy in WDM: Techniques and standards for component characterization. *Optical Fiber Communication Conference (OFC)*, Invited paper ThC1.

Glasser, L. A. (1980). A linearized theory for the diode laser in an external cavity, *IEEE J. Quantum Electron.* **QE-16**, 5, pp. 525-531.

Goedgebuer, L. P., Gurib, S. and Pote, H. (1992). Single frequency tuning of an external cavity diode laser at 1500 nm wavelength, *IEEE J. Quantum Electron.* **QE-28**, 6, pp. 1414-1418.

Goldberg, L., Taylor, H. F., Dandridge, A., Weller, J. F. and Miles, R. O. (1982). Spectral characteristics of semiconductor lasers with feedback, *IEEE J. Quantum Electron.* **QE-18**, 4, pp. 555-563.

Goldberg, L. and Kliner, D. A. V. (1995). Deep-UV generation by frequency quadrupling of a high-power GaAlAs semiconductor-laser, *Opt. Lett.* **20**, 10, pp. 1145-1147.

Goldberg, L. and Kliner, D. A. V. (1995). Tunable UV generation at 286 nm by frequency tripling of a high-power mode-locked semiconductor laser, *Opt.*

Lett. **20**, 15, pp. 1640-1642.

Goldberg, L., McElhanon, R. W. and Burns, W. K. (1995). Blue light generation in a bulk periodically field poled LiNbO3, *Electron. Lett.* **31**, 18, pp. 1576-1577.

Green, P. E. (1993). *Fiber optic networks*, Prentice Hall, Englewood Cliffs.

Greiner, G., Boggs, B., Wang, T. and Mossberg, T. W. (1998). Laser frequency stabilization by means of optical self-heterodyne beat-frequency control, *Opt. Lett.* **23**, 16, pp. 1280-1282.

Hall, R. N., Fenner, G. E., Kingsley, J. D., Soltys, T. J. and Carlson, R. O. (1962). Coherent light emission from GaAs junctions, *Phys. Rev. Lett.* **9**, (9), pp. 366-368.

Hansch, T. W. and Schawlow, A. L. (1975). Cooling of gases by laser radiation, *Optics Commun.* **13**, pp. 68.

Hansch T. W. and Couilaud, B. (1980). Laser frequency stabilization by polarization spectroscopy of a reflecting reference cavity, *Optics Commun.* **35**, pp. 441.

Harris, S. E. and Wallace, R. W. (1969). Acoustooptic tunable filter, *J. Opt. Soc. Am.* **59**, pp. 744.

Harris, S. E. (1992). Electromagnetically induced transparency, *Phys. Today.* **31**, pp. 711-715.

Harris, J. S. Jr,. (2000). Tunable long-wavelength vertical-cavity lasers: the engine of next generation optical networks?, *IEEE J. Sel. Top. Quantum Electron.* **6**, 6, pp. 1145-1160.

Harrison, J. and Mooradian, A. (1989). Linewidth and offset frequency locking of external cavity GaAlAs lasers, *IEEE J. Quantum Electron.* **25**, 6, pp. 1152-1155.

Harrison, P. (1999). *Quantum wells, wires, and dots, theoretical and computational physics*, Wiley.

Harvey, K. C. and Myatt, C. J. (1991). External-cavity diode laser using a grazing-incidence diffraction grating, *Opt. Lett.* **16**, 12, pp. 910-912.

Haus, H. A. (1980). Theory of modelocking of a laser diode in an external resonator, *J. Appl. Phys.* **51**, 18, pp. 4042-4049.

Hawthorn, C. J., Weber, K. P., Scholten, R. E. (2001). Littrow configuration tunable external cavity diode laser with fixed direction output beam, *Rev. Sci. Instrum.* **72**, 12, pp. 4477-4479.

Hayasaka, K. (2002). Frequency stabilization of an extended-cavity violet diode laser by resonant optical feedback, *Opt. Commun.* **206**, pp. 401-409.

Hayasaka, K. and Uetake, S. (2002). *XVIII International Conference on Atomic Physics,* Cambridge, Massachusetts, USA, pp. 146.

Hayashi, I., Panish, M. B., Foy, P. W. and Sumski, S. (1970). Junction lasers which operate continuously at room temperature, *Appl. Phys. Lett.* **17**, 3, pp. 109-111.

Hechscher, H. and Rossi, J. A. (1975). Flashing-size external-cavity semiconductor laser with narrow-linewidth tunable output, *Appl. Opt.* **14**, 1, pp. 94-96.

Heim, P. J. S., Fan, Z. F., Cho, S. H., Nam, K., Dagenais, M., Johnson, F. G. and Leavitt, R. (1997). Single-angled-facet laser diode for widely tunable external cavity semiconductor lasers with high spectral purity, *Electron. Lett.* **33**, 16, pp. 1387-1389.

Heismann, F., Alferness, R. C., Buhl, L. L., Eisenstein, G., Korotky, S. K., Veselka, J. J., Stulz, L. W. and Burrus, C. A. (1987). Narrow-linewidth elecrooptically tunable InGaAsP-Ti: LiNbo$_3$ extended cavity laser, *Appl. Phys. Lett.* **51**, 3, pp. 164-166.

Hemmerich, A., McIntyre, D. H., Schropp, Jr., D., Meschede, D. and Hansch, T. W. (1990). Optically stabilized narrow linewidth semiconductor laser for high resolution spectroscopy, *Opt. Commun.* **75**, 2, pp. 118-122.

Hemmerich, A., McIntyre, D. H., Zimmermann, C. and Hansch, T. W. (1990). Second-harmonic generation and optical stabilization of a diode laser in an external ring resonator, *Opt. Lett.* **15**, 7, pp. 372-375.

Henry, C. H. (1986). Phase niose in semiconductor lasers, *J. Lightwave Technol.* **LT-4**, pp. 293.

Hidaka, T. and Nakamoto, T. (1989). Electric tuning of semiconductor laser using acousto-optic device, *Electron. Lett.* **25**, 19, pp. 1320-1321.

Hildebrand, O., Schilling, M., Baums, D., Idler, W., Dutting, K., Laube, G. and Wunstel, K. (1993). The Y-laser : A multifunctional device for optical communication systems and switching networks, *J. Lightwave Technol.* **11**, 12, pp. 2066-2075.

Hildebrandt, L., Knispel. R., Stry, S., Sacher, J. R. and Schael, F. (2003). Antireflection-coated blue GaN laser diodes in an external cavity and Doppler-free indium absorption spectroscopy, *Appl. Opt.* **42**, 12, pp. 1-9.

Hirota, O. and Suematsu, Y. (1979). Noise properties of injection lasers due to reflected waves, *IEEE J. Quantum Electron.* **QE-15**, 3, pp. 142-149.

Hjelme, A. R. and Mickelson, A. R. (1987). On the theory of external-cavity controlled semiconductor lasers, *IEEE J. Quantum Electron.* **QE-23**, 6, pp. 1000-1004.

Ho, P. T., Glasser, L. A., Ippen E. P. and Haus, H. A. (1978). Picosecond pulse geberation with a cw GaAlAs laser diode, *Appl. Phys. Lett.* **33**, 3, pp. 241-242.

Hof, T., Fick, D. and Jansch, H. J. (1996). Application of diode laers as a spectroscopic too at 670 nm, *Opt. Commun.* **124**, pp. 283-286.

Holbrook, M. B., Sleat, W. E. and Bradley, D. J. (1980). Bandwidth-limited picosecond pulse generation in an actively mode-locked GaAlAs diode laser, *Appl. Phys. Lett.* **37**, 1, pp. 59-61.

Hollberg, L. and Ohtsu, M. (1988). Modulatable narrow-linewidth semiconductor lasers, *Appl. Phys. Lett.* **53**, 11, pp. 944-946.

Holonyak, Jr., N. and Bevacqua, S. F. (1962). Coherent (visible) light emission from Ga ($As_{1-x}P_x$) junctions, *Appl. Phys. Lett.* **1**, 4, pp. 82-83.

Hori, H. et al. (1983). Frequency stabilization of GaAs laser using a Doppler-free spectrum of the Cs-D$_2$ line, *IEEE. J. Quantum Electron.* **QE-19**, pp. 169.

Hui, R. Q. and Tao, S. P. (1989). Improved rate equation for external-cavity semiconductor lasers, *IEEE J. Quantum Electron.* **QE-25**, 6, pp. 1580-1584.

Ikegami, T., Sudo, S. and Sakai, Y. (1995). *Frequency stabilization of semicon-*

ductor laser diodes. Artech House, Boston.

Inguscio, M. and Wallenstain, R. (1993). *Solid State Lasers, New developments and applications*, Plenum Press, New York.

Inguscio, M. (1994). High-resolution and high-sensitivity spectroscopy using semiconductor didoe lasers, in *Proc. of the international school of physics, Frontier in laser spectrocopy,* edited by T. W. Hansch and M. Inguscio, North-Holland.

Ishida, O. and Toba, H. (1991). 200 kHz absolute grequency stability in 1.5 μm external-cavity semiconductor laser, *Electron. Lett.* **27**, 12, pp. 1018-1019.

Ishii, H., Kano, F., Tohmori, Y., Kondo, Y., Tamaruma, T. and Yoshikuni, Y. (1994). Broad range (34 nm) qusi-continuous wavelength tuning in superstructure-grating DBR lasers, *Electron. Lett.* **30**, 14, pp. 1134-1135 .

Ishii, H., Tanobe, H., Kano, F., Tohmori, Y., Kondo, Y. and Yoshikuni, Y. (1996). Qusidiscontinuous wavelength tuning in a super-structure-grating (SSG) DBR lasers, *IEEE J. Quantum Electron.* **QE-32**, 3, pp. 433-441.

Ishii, H., Tanobe, H., Kano, F., Tohmori, Y., Kondo, Y. and Yoshikuni, Y. (1996). Broad-range wavelength coverage (62.4 nm) with superstructure-grating DBR lasers,*Electron. Lett.* **32**, 5, pp. 454-455.

Jayaraman, V., Cohen, D. A. and Coldren, L. A. (1992). Demonstration of broadband tunability in a semiconductor laser using sampled grating, *Appl. Phys. Lett.* **60**, 19, pp. 2321-2323.

Jayaraman, V., Chuang, Z. M. and Coldren, L. A. (1993). Theory, design, and performance of extended tuning range semiconductor lasers with sampled gratings, *IEEE J. Quantum Electron.* **QE-29**, 6, pp. 1824.

Jones, R. J., Gupta, S., Jain, R. K. and Walpole, J. N. (1995). Near diffraction-limited high power (1 W) single longitudal mode CW diode laser tunable from 960 nm to 980 nm, *Electron. Lett.* **31**, 19, pp. 1668-1669.

Kapon, E. (1999). *Semiconductor lasers I & II*, Ed. Academic Press, San Diego.

Karlsson, C. J. and Olsson, F. A. A. (1999). Linearization of the frequency sweep of a frequency-modulated continuous-wave semiconductor laser radar and the resulting ranging performance, *Appl. Opt.* **38**, pp. 3376.

Kiang, M. H., Solgaad, O., Muller, R. S. and Lau, K. Y. (1996). Silicon-micromachined micromirrors with integrated high-precision actuators for external-cavity semiconductor lasers, *IEEE Photo. Tech. Lett.* **8**, 1, pp. 95-98.

Kim, I., Alferness, R. C., Koren, U., Buhl, L. L., Miller, B. I., Young, M. G., Chien, M. D., Koch, T. L., Presby, H. M., Raybon, G. and Burrus, C. (1994). Broadly tunable vertical coupler filter tensile stained InGaAsP/InP multiple quantum well laser, *Appl. Phys. Lett.* **64**, 21, pp. 2764-2766.

Kintzer, E. S., Walpole, J. N., Chinn, S. R., Wang, C. A. and Missaggia, L. J. (1993). High-pwer strained-layer amplifier and lasers with tappered gain regions, *IEEE Photo. Tech. Lett.* **5**, pp. 605-607.

Kitamura, K., Yamaguchi, M., Emura, D., Mito, I. and Kobayashi, K. (1985). Lasing mode and spectral linewidth control by phase tunable distributed feedback laser diodes with double channel planar buried heterostructure (DFB-DC-PBH), *IEEE J. Quantum Electron.* **QE-21**, 5, pp. 415-417.

Kitching, J., Knappe, S., Vukicevic, N., Hollberg, L., Wynands, R. and Weidmann, W. (2001). A microwave frequency reference based on VCSEL-driven dark line resonances in Cs vapor, *IEEE Trans. Instru. Meas.* **49**, 6, pp. 1313-1317.

Kitching, J., Robinson, H. G., Hollberg, L., Knappe, S. and Wynands, R. (2001). Optical-pumping niose in laser-pumped, all-optical microwave frequency references, *J. Opt. Soc. Am. B* **18**, 11, pp. 1676-1683.

Kittle, C. (1982). *Introduction to solid state physics*, 5th, New York: Wiley.

Kliner, D. A. V., Koplow, J. P. and Goldberg, L. (1997). Narrow-band, tunable semiconductor-laser-based source for deep-UV absorption spectroscopy, *Opt. Lett.* **22**, 18, pp. 1418-1420.

Kner, P., Kageyama, T., Boucart, J., Stone, R., Sun, D., Nabiev, R. F., Pathak, R. and Yuen, W. (2003). A long-wavelength MEMS tunable VCSEL incorporating a tunnel junction, *IEEE Photon. Technol. Lett.* **15**, 9, pp. 1183-1185.

Kobayashi, K. and Mito, I. (1988). Single frequency and tunable laser diodes, *J. Lightwave Tech.* **6**, 11, pp. 1623-1633.

Koch, T. L. and Koren, U. (1990). Semiconductor lasers for coherent optical fiber communications, *J. Lightwave Tech.* **8**, 3, pp. 274-293

Kogelnik, H. and Shank, C. V. (1972). Stimulated emission in a periodic structure, *Appl. Phys. Lett.* **18**, 4, pp. 152-154.

Koplow, J., Kliner, D. A. V. and Goldberg, L. (1998). UV generation by frequency quadrupling of a Yb-doped fiber amplifier, *IEEE Photon. Technol. Lett.* **10**, 1, pp. 75-77.

Kourogi, M., Imai, K., Widyatmoko, B., Shimizu, T. and Ohtsu, M. (2000). Continuous tuning of an electrically tunable external cavity semiconductor laser, *Opt. Lett.* **25**, 16, pp. 1165-1167.

Kowalski, F. V., Nakamura, K. and Ito, H. (1998). Frequency shifted feedback lasers: continuous or stepwise frequency chirped output? *Opt. Commun.* **147**, pp. 103-106.

Kovlovsky, W. J., and Lenth, W. (1990). Generation of 41 mW of blue radiation by frequency doubling of a GaAlAs diode laser, *Appl. Phys. Lett.* **56**, pp. 2291-2293.

Kroll, S. and Elman, U. (1993). Photon-echo-based logical processing, *Opt. Lett.* **18**, pp. 1834-1836.

Kruger, J. M. W. (1998). *A novel technique for frequency stabilizing laser diodes*, Thesis of Univeristy of Otago.

Kudo, K., Yashiki, K., Sasaki, T., Yokoyama, Y., Hamamoto, K., Morimoto, T. and Yamaguchi, M. (2000). 1.55μm wavelength-selectable microarray DFB-LD's with monolithically integrated MMI combiner, SOA, and EA-Modulator, *IEEE Photon. Technol. Lett.* **12**, 3, pp. 242-244.

Kuhl, J., Serenyi, M. and Gobel, E. O. (1987). Bandwidth-limited picosecond pulse generation in an actively mode-locked GaAs laser with intracavity chirp compensation, *Opt. Lett.* **12**, 5, pp. 334-336.

Kuznetsov, M. (1988). Theory of wavelength tuning in two-segment distributed feedback lasers, *IEEE J. Quantum Electron.* **QE-24**, 9, pp. 1837-1844.

Ladany, I., Ettenberg, M., Lockwood, H. F. and Kressel, H. (1977). *Al_2O_3 half-wave films for long-life cw lasers*, *Appl. Phys. Lett.* **30**, 2, pp. 87-88.

Lancaster, G. P. T., Sibbett, W. and Dholakia, K. (2000). An extended-cavity diode laser with a circular output beam, *Rev. Sci. Instrum.* **71**, 10, pp. 3646-3647.

Lang, R. and Kobayashi, R. (1980). External optical feedback effects on semiconductor injection laser properties, *IEEE J. Quantum Electron.* **QE-16**, 3, pp. 347-355.

Lang, R. J., Yariv, A. and Salzman, J. (1987). Laterally coupled-cavity semiconductor laser, *IEEE J. Quantum Electron.* **QE-23**, 4, pp. 395-400.

Lang, M. (1998). External-cavity diode lasers provide absolute reference for WDM testing, *Laser Focus World*, June.

Langley, L. N., Shore, K. A. and Mork, J. (1994). Dynamical and noise properties of laser diodes subject to strong feedback, *Opt. Lett.* **19**, 24, pp. 2137-2139.

Langley, L. N., Turovets, S. and Shore, K. A. (1995). Targeting periodic oscillations of external cavity laser diodes, *Opt. Lett.* **20**, 7, pp. 725-727.

Larson, M. C., Pezeshki, B. and Harris, Jr., J. S. (1995). Vertical coupled-cavity microinterferometer on GaAs with deformable-membrane top mirror, *IEEE Photon. Technol. Lett.* **7**, 4, pp. 382-384.

Larson, M. C., and Harris, Jr., J. S. (1996). Wide and continuous wavelength tuning in a vertical-cavity surface-emitting laser using a micormachined deformable-membrance mirror, *Appl. Phys. lett.* **68**, 7, pp. 891-893.

Larson, M. C., Kondow, M., Kitatani, T., Nakahara, K., Tamura, K., Inoue, H. and Uomi, K. (1998). GaInNAs-GaAs long-wavelength Vertical-cavity surface-emitting laser diodes, *IEEE Photon. Technol. Lett.* **10**, 2, pp. 188-190.

Lau, L. Y. (1991). Narrow linewidth, continuously tunable semiconductor lasers based on quantum well gain level, *Appl. Phys. Lett.* **59**, 18, pp. 2216-2218.

Laude, J. P. (1993). *Wavelength division multiplexing,* Prentice Hall, New York.

Laurent, P., Clairon, A. and Breant, C. (1989). Frequency noise analysis optically self-locked diode lasers, *IEEE J. Quantum Electron.* **25**, 6, pp. 1131-1142.

Lazar, J., Jedlicka, P. Cip, O. and Ruzicka, B. (2003). Laser diode current controller with a high level of protection against electromagnetic interference, *Rev. Sci. Instrum.* **74**, 8, pp. 3816-3819.

Leinen, H., Glaβner, D., Metcalf, H., Wynands, R., Haubrich, D. and Meschede, D. (2000). GaN blue didoe lasers: a spectroscopists's view, *Appl. Phys. B* **70**, pp. 567-571.

Lenstra, D., Verbeek, B. H. and den Boef, A. J. (1985). Coherence collapse in single-mode semiconductor lasers due to optical feedback, *IEEE J. Quantum Electron.* **21**, 6, pp. 674-679.

Levin, L. (2002). Mode-hop-free electron-optically tuned didoe laser, *Opt. Lett.* **27**, 4, pp. 237-239.

Li, H. and Telle, H. R. (1989). Efficient frequency noise reduction of GaAlAs semiconductor lasers by optical feedback from an external high-finesse resonator, *IEEE J. Quantum Electron.* **25**, 3, pp. 257-264.

Li, M., Yuen, W., Li, G. S. and Chang-Hasnain, C. J. (1997). High performance continuously tunable top-emitting micromechanic vertical cavity surface emittig lasers with 20 nm wavelength range, *Electron. Lett.* **33**, 12, pp. 1051-1052.

Li, M., Yuen, W., Li, G. S. and Chang-Hasnain, C. J. (1998). Top-emitting micromechanic vertical cavity surface emittig lasers with a 31.6 nm tuning range, *IEEE Photo. Tech. Lett.* **10**, 1, pp. 18-20.

Li, H., Liu, G. T., Varangis, P. M., Newell, T. C., Stintz, A., Fuchs, B., Malloy, K. J. and Lester, L. F. (2000). 150 nm tuning range in a grating-coupled external cavity quantum-dot laser, *IEEE Photo. Tech. Lett.* **12**, 7, 759-761.

Li, H. and Iga, K. (2003). *Vertical-cavity surface-emitting laser devices*, Springer.

Libbrecht, K. G., and Hall, J. L. (1993). A low-noise high-speed diode laser current controller, *Rev. Sci. Instrum.* **64**, 8, pp. 2133-2135.

Libbrecht, K. G., Boyd, R. A., Willems, P. A., Gustavson, T. L. and Kim, D. K. (1995). Teaching physics with 670 nm didode lasers -construction of stabilized lasers and lithium cells, *Am. J. Phys.* **63**, 8, pp. 729-737.

Lin, H., Wang, T., Wilson, G. A. and Mossberg, T. W. (1995). Experimental demonstration of swept-carrier time-domain spatial memory, *Opt. Lett.* **20**, pp. 91-93.

Littman, M. G. (1978). Single-mode operation of grazing-incidence pulsed dye laser, *Opt. Lett.* **3**, 4, pp. 138-141.

Littman, M. G. and Metcalf, H. J. (1978). Spectrally narrow pulsed dye laser without beam expander, *Appl. Opt.* **17**, 14, pp. 2224-2227.

Littman, M. G. (1984). Single-mode pulsed tunable dye laser, *Appl. Opt.* **23**, 24, pp. 4465-4468

Liu, K. and Littman, M. G. (1981). Novel geometry for single-mode scanning of tunable lasers, *Opt. Lett.* **6**, 3, pp. 117-118.

Liu, P. L., Eisenstein, G., Tucker, R. S. and Kaminow, I. P. (1984). Measurements of intensity fluctuations of an InGaAsP external cavity laser, *Appl. Phys. Lett.* **44**, 51, pp. 481-483.

Liu, B., Shakouri, A. and Bowers, J. E. (2002). Wide tunable double ring resonator coupled lasers, *IEEE Photo. Tech. Lett.* **14**, 5, pp. 600-102.

Liu A. Q., Zhang, X. M., Murukeshan, V. M., Lu, C. and Cheng, T. H. (2002). Micromachined wavelength tunable laser with an extended feedback model, *IEEE J. Select. Topics Quantum electron.* **8**, 1, pp. 73-79.

Liu, H. F., Oshiba, S., Ogawa, Y. and Kawai, Y. (1992). Method of generating nearly transform-limited pulses from gain-switched distributed-feedback laser diodes and its application to soliton transmission, *Opt. lett.* **17**, 1, pp. 64-66.

Lohmann, A. and Syms, R. R. A. (2003). External cavity lasers with a vertically etched silicon blazed grating, *IEEE Photo. Tech. Lett.* **15**, 1, pp. 120-122.

Lotem, H., Pan, Z, and Dagenais, M. (1992). Tunable external cavity diode laser that incorporates a polarization half-wave plate, *Appl. Opt.* **31**, 36, pp. 7530.

Loyt, B. (1933). Optical apparatus with wide field using interference of polarized light, *C. R. Acad. Sci.*(Paris) **197**, pp. 1593.

Ludeke, R., and Harris, E. P. (1972). Tunable GaAs laser in an external cavity dispersive cavity, *Appl. Phys. Lett.* **20**, 12, pp. 499-500.

MacAdam, K. B., Steinbach, A., Wieman, C. E. (1992). A narrow-band tunable laser system with grating feedback, and a saturated absorption spectrometer for Cs and Rb, *Am. J. Phys.* **60**, 12, pp. 1098-1111.

Maki, J. J., Campbell, N. S., Grande, C. M., Knorpp, R. P. and McIntyre, D. H. (1993). Stabilized diode-laser system with grating feedback and frequency-offset locking, *Opt. Commun.* **102**, pp. 251-256.

Mason, B., Lee, S. L., Heimbuch, M. E. and Coldren, L. A. (1997). Directly modulated sampled grating DBR lasers for long-haul WDM communication systems, *IEEE J. Photon. Technol. Lett.* **9**, 3, pp. 377-379.

Mason, B., Fish, G. A., Barton, J., Kaman, V., Coldren, L. A., Denbaars, S. P. and Bowers, J. (2000). Characteristics of smapling grating DBR lasers with integrated semoconductor optical amplifiers and electron absorption modulators, in *Proc. OFC.*

Mcnicholl, P. and Metcalf, H. J. (1985). *Appl. Opt.* **24**, pp. 2757.

Mehuys, D., Mittelstein, M., Yariv, A., Sarfaty, R. and Ungar, J. E. (1989). Optimised Fabry-Perot AlGaAs quantum-well lasers tunable over 105 nm, *Electron. Lett.* **25**, 2, pp. 143-145.

Mehuys, D., Welch, D. and Scifres, D. (1997). 1 W CW diffraction-limited tunable external-cavity semiconductor laser, *Electron. Lett.* **29**, 14, pp. 1254-1255.

Mellis, J., Al-chalabi, S. A., Cameron, K. H., Wyatt, R., Regnault, J. C., Delvin, W. J. and Brain, M. C. (1988). Miniature packaged external-cavity semiconductor laser with 50 GHz continuos electrical tuning range, *Electron. Lett.* **24**, 16, pp. 988-989.

Mnager, L., Cabaret, L., Lorger, I. and Le Gout, J. L. (2000). Diode laser extended cavity for broad-range fast ramping, *Opt. Lett.* **25**, 17, pp. 1246-1248.

L. Mnager, L. Cabaret, I. Lorger, J.-L. Le Gout

Merritt, S. A., Dauga, C., Fox, S., Wu, I. F. and Dagenais, M. (1995). Measurement of the facet modal reflectivity spectrum in high quality semiconductor traveling wave amplifiers, *J. Lightwave Technol.* **13**, 3, pp. 430-433.

Merkel, K. D. and Babbitt, W. R. (1996). Coherent transient optical signal processing without brief pulses, *Appl. Opt.* **35**, pp. 278.

Meziane, B., Besnard, P. and Stephan, G. M. (1995). Low-frequency resonances in asymmetric external cavity semiconductor lasers: theory and experiment, *IEEE J. Quantum Electron.* **31**, 4, pp. 617-622.

Milic, D., Lu, W., Hoogerland, M. D., Blacksell, M., Baldwin, K. G. H. and Buckman, S. J. (1997). Improved spectral properties of diode lasers, *Rev. Sci. Instrum.* **68**, 10, pp. 3657-3659.

Mittelstein, M., Mehuys, D., Yariv, A., Ungar, J. E. and Sarfaty, R. (1989). Broadband tunibility of gain-flattened quantum well semiconductor laser with an external grating, *Appl. Phys. Lett.* **54**, 12, pp. 1092-1094.

Morgott, S., Chazan, P., Mikulla, M., Walther, M., Kiefer, R., Braunstein, J. and Weimann, J. (1988). High-power near diffraction-limited external cavity, tunable from 1030 nm to 1085 nm, *Electron. Lett.* **34**, 6, pp. 558-559.

Mork, J., Tromborg, B. and Christiansen, P. L. (1988). Bistability and low frequency fluctuation in semiconductor lasers with optical feedback: A theoretical analysis, *IEEE J. Quantum Electron.* **24**, 2, pp. 123-133.

Morton, P. A., Mizrahi, V., Kosinski, S. G., Mollenauer, L. F., Tanbun-Ek, T., Logan, R. A., Coblentz, D. L., Sergent, A. M. and Wecht, K. W. (1992). Hybrid soliton pulse source with fibre external cavity and Bragg reflector, *Electron. Lett.* **28**, 6, pp. 561-562.

Morzinski, J., Bhatia, P. S. and Shahriar, M. S. (2002). Frequency stabilizaiton of an external cavity semiconductor diode laser for chirping, Rev. Sci. Instrum. **73**, 10, pp. 3449-3453.

Murata, S., Mito, I. and Kobayashi, K. (1987). Frequency modulation and spectral characteristics for a 1.5 μm phase-tunable DFB lasers, *Electron. Lett.* **23**, pp. 12-13.

Myatt, C. J., Newbury, N. R. and Wieman, C. E. (1993). Simplified atom trap by using direct microwave modulation of a diode lasers, *Opt. Lett.* **18**, 8, pp. 649-651.

Nagel, A., Graf, L., Naumov, A., Mariotti, E., Biancalana, V., Meschede, D. and Wynands, R. (1998). *Europhys. Lett.* **44**, 1, pp. 31-36.

Nakamura, M. S. and Ohshima, S. (1990). Frequency-stabilized LD module with a Z-cut quarta Febry-Perot resonator for coherent communication, *Electron. Lett.* **26**, pp. 405-406.

Nakamura, S. and Fasol, G. (1997). *The blue laser diode*, Springer.

Nakamura, K., Kowalski, F. V. and Ito, H. (1997). Chirped-frequency generation in a translated-grating-type frequency-shifted feedback laser, *Opt. Lett.* **22** 12, pp. 889-891.

Nakamura, K., Miyahara, T., Yoshida, M., Hara, T. and Ito, H. *IEEE. Photonics Technol. Lett.* **10**, 1772, (1998).

Nakamura, K., Miyahara, T. and Ito, H. (1998). Observation of a highly phase-correlated chirped frequency comb output from a frequency-shifted feedback laser, *Appl. Phys. Lett.* **72**, 12, pp. 2631-2633.

Nathan, M. I., Dumke, W. P., Burns, G., Dill, Jr., F. H. and Lasher, G. J. (1962). Stimulated emission of radiation from GaAs p-n junctions, *Appl. Phys. Lett.* **1**, 3, pp. 62-64.

Novikova, I., Matsko, A. B., Velichansky, V. L., Scully, M. O. and Welch, G. R. (2001). Compensation for ac Stark shifts in optical magnetometry, *Phys. Rev. A* 63, pp. 063802.

Oberg, M., Nilsson, S., Klinga, T. and Ojala, P. (1991). *IEEE J. Photon. Technol. Lett.* **3**, pp. 299-301.

Oberg, M., Nilsson, S., Strerbel, K., Wallin, J., Backbom, L. and Klinga, T. (1993). 74 nm wavelength tuning range of an InGaAsP/InP vertical grating assisted codirectional coupler laser with rear sampled grating reflector, *IEEE J. Photon. Technol. Lett.* **5**, 7, pp. 735-738.

Ogorman, G., and Levi, A. F. J. (1993). Wavelength dependence of T_0 in InGaAsP semiconductor laser diodes, *Appl. Phys. Lett.* **62**, 17, pp. 2009-2011.

Oh, D. B. (1995). Diode-laser-based sum-frequency generation of tunable wavelength-modulated UV light for OH radical detection, *Opt. Lett.* **20**, 1, pp. 100-102.

Ohtsu, M. (1991). *Highly coherent semiconductor lasers*, Artech House.

Okoshi, T and Kikuchi, K. (1981). Heterodyne-type optical fiber communications, *J. Opt. Commun.* **2**, pp. 82-88.

Okuda, M. and Onaka, K. (1977). Tunability of distributed Bragg-reflector laser by modulating refractive index incorrugated waveguide, *Japanese J. Appl. Phys.* **16**, pp. 1501.

Olsson, A. and Tang, C. L. (1981). Coherent optical interference effects in external cavity semiconductor lasers, *IEEE J. Quantum Electron.* **QE-17**, 8, pp. 1320-1323.

Olsson, N. A., Henry, C. H., Kazarinov, R. F., Lee, H. J., Johnson, B. H. and Orlowsky, K. J. (1987). Narrow linewidth 1.5 μm semiconductor laser with a resonant optical reflector, *Appl. Phys. Lett.* **51**, 15, pp. 1141-1142.

Osmundsen, J. H. and Gade, N. (1983). Influence of optical feedback on laser frequency spectrum and threshold conditions, *IEEE J. Quantum Electron.* **QE-19**, 3, pp. 465-469.

Osinski, M. and Buus, J. (1987). Linewidth broadening factor in semiconductor laser, an overreview, *IEEE J. Quantum Electron.* **23**, 1, pp. 9-29.

Palmer, C. (2002). *Diffraction Gratings Handbook*, 5th edition, Thermal RGL.

Pan, X., Olessen, H. and Tromborg, B. (1988). A theoretical model of multielectrode DBR lasers, *IEEE J. Quantum Electron.* **24**, 12, pp. 2423-2432.

Park, J. D., Seo, D. S. and McInerney, J. G. (1990). Self-pulsations in strongly asymmetric external semiconductor lasers, *IEEE J. Quantum Electron.* **26**, 8, pp. 1353-1362.

Petermann, K. (1988). *Laser diode modulation and noise*, London, U.K. Kluwer Academic Publishers.

Peterson, K. A. and Oh, D. B. (1999). High-sensitivity detector of CH radicals in flames by use of a diode-laser-based near-ultra light source, *Opt. Lett.* **24**, 10, pp. 667-669.

Petridis, C., Lindsay, I. D., Stothard, D. J. M. and Ebrahimzadeh, M. (2001). Mode-hop-free tuning over 80 GHz of an external cavity diode laser without antireflection coating, *Rev. Sci. Instru.* **72**, 10, pp. 3811-3815.

Petuchowski, S. J., Miles, R. O., Dandridge, A. and Giallorenzi, T. D. (1982). Phase sensitivity and linewidth narrowing in a Fox-Smith configured semiconductor laser, *Appl. Phys. Lett.* **40**, 4, pp. 302-304.

Pezeshki, B., Mathur, A., Zhou, S., Jeon, H. S., Agrawal, V. and Lang, R. L. (2000). 12 nm tunable WDM source using an integrated laser array, *Electron. Lett.* **36**, 9, pp. 788-789.

Pezeshki, B., Vail, E., Cubicky, J., Yoffe, G. Zou S. et al. (2002). 20 mW widely tunable laser module using DFB array and MEMS selection, *IEEE Photon. Tech. Lett.* **14**, 10, pp. 1457-1459.

Picque, J. L. and Roison, S. (1975). Frequency-controlled cw tunable GaAs laser, *Appl. Phys. Lett.* **27**, 6, pp. 340-342.

Poelker, M., Kumar, P. and Hoe, S. T. (1991). Laser frequenction transition: a new method, *Opt. Lett.* **16**, 23, pp. 1853-1855.

Pound, R. V. (1946). Electronic frequency stabilization of microwave oscillators, *Rev. Sci. Instrum.* **17**, pp. 490.

Quist, T. M., Rediker, R. H., Keyes, R. J., Krag, W. E., Lax, B., McWhorter, A. L. and Zeiger, H. J. (1962). *Appl. Phys. Lett.* **1**, pp. 91.

Rapasky, K. S., Switzer, G. W. and Carlsten J. L. (2002). Design and performance of a frequency chirped external cavity diode laser, *Rev. Sci. Instrum.* **73**, 9, pp. 3154-3159.

Ray, G. J., Anderson, T. N., Caton, J. A., Lucht, R. P. and Walther, Th. (2001). OH sensor based on ultraviolet, continuous-wave avsorption spectroscopy utilizing a frequency-quadrupled, fiber-amplified external-cavity diode laser, *Opt. Lett.* **26**, 23, pp. 1870-1872.

Reithmaier, J. R. and Forchel, A. (2002). Single-mode distributed feed and microlasers based on quantum-dot gain material, *IEEE J. Sel. Top. Quantum Electron.* **8**, 5, pp. 1035-1044.

Ricci, L., Weidemuller, M., Esslinger, T., Hemmmerich, A., Zimmermann, C., Vuletic, V., Konig, W. and Hansch, T. W. (1995). A compact grating-stabilized diode laser system for atomic physics, *Opt. Commun.* **117**, pp. 541-549

Rideout, W., Holestrom, R., Lacourse, J., Meland, E. and Powazinik, W. (1990). Ultra-low-reflectivity semiconductor optical amplifiers without antireflection coating, *Electron. Lett.* **26**, 1, pp. 36-38.

Rigole, P. J., Nilsson, S., Backbom, L., Klinga, T., Wallin, J., Stalnacke, B., Berglind, E. and Stoltz, B. (1995). 114 nm wavelength tuning range of a vertical grating assisted codirectional coupler laser with a super structure grating distributed Bragg reflector, *IEEE J. Photon. Technol. Lett.* **7**, 7, pp. 697-699.

Rigole, P. J., Nilsson, S., Backbom, L., Klinga, T., Wallin, J., Stalnacke, B., Berglind, E. and Stoltz, B. (1995). Access to 20 evenly distributed wavelengths over 100 nm using only a single current tuning in a four-electrode monolithic semiconductor laser, *IEEE J. Photon. Technol. Lett.* **7**, 11, pp. 1249-1251.

Ross, S. B., Kanorsky, S. I., Weis, A. and Hansch, T. W. (1995). A single mode cw diode lase at the cesium D_1 (894.59 nm), *Opt. Commun.* **120**, pp. 155-157.

Rovera, G. D., Santarelli, G. and Clairon, A. (1994). A laser diode system stabilized on the cesium D_2 line, *Rev. Sci. Instrum.* **65**, 5, pp. 1502-1505.

Sacher, J., Elsaβer, W. and Gobel, E. O. (1989). Intemittency in the coherence collapse of a semiconductor lasers with external feedback, *Phys. Rev. Lett.* **63**, 20, pp. 2224-2227.

Salathe, R. P. (1979). Diode lasers coupled to external resonators, *Appl. Phys.* **20**, pp. 1-18.

Salomon C., Dalibard, J., Aspect, A., Metcalf, H. and Cohen-Tannoudji, C. (1987). Channeling atoms in a laser standing wave, *Phys. Rev. Lett.* **59**, 15, pp. 1659-1662.

Sato, T., Niikuni, M., Sato, S. and Shimba, M. (1988). Frequency stabilization of a semiconductor laser using a Rb-D_1 and D_2 absorption lines, *Electron. Lett.* **24**, 7, pp. 429-431.

Sayama, S. and Ohtsu, M. (1998). Tunable UV cw generation at 276 nm wavelength by frequency conversion of laser diodes, *Opt. Commun.* **145**, pp.

95-97.

Schremer, A. T. and Tang, C. L. (1990). External-cavity semiconductor laser with 1000 GHz continuous piezoelectric tuning range, *IEEE Photo. Tech. Lett.* **2**, 1, pp. 3-5.

Schell, M., Weber, A. G., Scholl, E. and Bimberg, D. (1991). Fundamnetal limits of sub-ps pulse generation by active mode locking of semiconductor lasers: the spectral gain width and the facet reflectivities, *IEEE J. Quantum Electron.* **QE-27**, 6, pp. 1661-1668.

Schunk, N. and Petermann, K. (1988). Numerical analysis of the feedback regimes for a single-mode semiconductor laser with external feedback. *IEEE J. Quantum Electron.* **QE-24**, 7, pp. 1242-1247.

Scully, M. O. and Fleischhauer, M. (1992). High-sensitivity magnetometer based on index-enhanced media, *Phys. Rev. Lett.* **69**, 9, pp. 1360-1363.

Sellin, P. B., Strickland, N. M., Carlsten, J. L. and Cone, R. L. (1999). Programmable frequency reference for subkilohertz laser stabilization by use of persistent spectral hole burning, *Opt. Lett.* **24**, 15, pp. 1038-1040.

Seo, D. S., Park, J. D., McInerney, J. G. and Osinski, M. (1988). Effects of feedback asymmetry in external-cavity semiconductor laser systems, *Electron. Lett.* **24**, 12, pp. 727-728.

Shi, H. X., Cohen, D., Barton, J., Majewski, M., Coldren, L. A., Larson, M. C. and Fish, G. A. (2002). Relative intensity noise measurements of a widely tunable sampled-grating DBR laser, *IEEE Photo Tech. Lett.* **14**, 6, pp. 759-761.

Shoshan, I., Danon, N. N. and Oppenheim, U. P. (1977). Narrowband operation of a pulsed dye laser without intracavity beam expansion, *J. Appl. Phys.* **48**, 11, pp. 4495-4497.

Shoshan, I. and Oppenheim, U. P. (1978). The use of a diffraction grating as a beam expander in a dye laser cavity, *Opt. Commun.* **25**, 3, pp. 375-378.

Siegman, A. E. (1986). *Lasers*, Mill valley, CA, Univeristy science, Chapters 27, and 28.

Sigg, J. (1993). Effects of optical feedback on the light-current characteristics of semiconductor lasers, *IEEE J. Quantum Electron.* **QE-29**, 5, pp. 1262-1270.

Sivaprakasam, S., Saha, R., Lakshmi, P. A. and Singh, R. (1996). Mode hopping in external-cavity diode laser, *Opt. Lett.* **21**, 6, pp. 411-413 (1996).

Snadden, M. J., Clarke, R. B. M. and Riis, E. (1997). Injection-locking technique for heterodyne optical phase locking of a diode laser, *Opt. Lett.* **22**, 12, pp. 892-894. (1996).

Struckmeier, J., Euteneuer, A., Smarsly, B., Breede, M., Born, M. and Hofmann, M. (1999). Electronically tunable external-cavity laser diode, *Opt. Lett.* **24**, 22, pp. 1573-1575.

Sugihwo, F., Larson, M. C. and Harris, Jr., J. S. (1998). Simultaneous optimization of membrane reflectance and tuning voltage for tunable vertical cavity laser, *Appl. Phys. lett.* **72**, 1, pp. 10-12.

Sun, X. G., Switzer, G. W. and Carlsten, J. L. (2000). Bule light generation in an external ring cavity using both cavity and grating feedback, *Appl. Phys. Lett.* **76**, 8, pp. 955-957.

Sun. D., Fan, W., Kner, P., Boucart, J., Kageyama, T., Zhang, D. X., Pathak, R., Nabiev, R. F. and Yuen, W. (2004). Long wavelength-tunable VCSELs with optimized MEMS bridge tuning structure, *IEEE Photo. Tech. Lett.* **16**, 3, pp. 714- 716.

Svelto, O. (1998). Translated and Edited by D. C. Hanna, *Principles of lasers*, 4th Edition, Plenum Press, NewYork.

Swann, W. C. and Gilbert, S. L. (2000). Pressure-induced shift and broadening of 1510~1540 nm acetylene wavelength calibration lines, *J. Opt. Soc. Am. B* **17**, pp. 1263.

Tabuchi, H. and Ishikawa, H. (1990). External grating tunable MQW laser with wide tuning range of 240 nm, *Electron Lett.* **26**, 11, pp. 742-743.

Takiguchi, Y., Liu, Y. and Ohtsubo, J. (1998). Low-frequency fluctuation induced by injection-current modulation in semiconductor lasers with optical feedback, *Opt. Lett.* **23**, 17, pp. 1369-1371.

Talvitie, H., Pietilainen, A., Ludvigsen, H. and Ikonen E. (1997). Passive frequency and intenisty stabilization of extended-cavity diode lasers, *Rev. Sci. Instrum.* **68**, 1, pp. 1-7.

Tang, C. L., Kreismanis, V. G. and Ballantyne, J. M. (1977). Wide-band electro-optical tuning of semiconductor lasers, *Appl. Phys. Lett.* **30**, 1, pp. 113-116.

Tanner, C. E. and Wieman, C. E. (1988). Precision measurement of the Stark effect in the $6S_{1/2} \rightarrow 6P_{3/2}$ cesium transition using a frequency-stabilized laser diode, *Phys. Rev. A* **38**, 1, pp. 162.

Taylor, D. J., Harris, S. E., Nich, S. T. K. and Hanch, T. W. (1971). Electrically tuning of a dye laser using the acousto-optic filter, *Appl. Phys. Lett.* **19**, pp. 269-270.

Thompson, G. H. B. and Kirby, P. A. (1973). (GaAl)As lasers with a heterostructure for optical confinement and additional heterojunctions for extreme carrier confinement, *IEEE J. Quantum Electron.* **QE-9**, 2, pp. 311-318.

Thompson, G . H. B. (1980). *Physics of semiconductor laser devices*, Wiley, New York.

Tohmori, Y., Yoshikuni, Y., Ishii, H., Kano, F., Tamaruma, T., Kondo, Y. and Yamamoto, M. (1993). Broad-range wavelength-tunable superstructure grating (SSG) DBR lasers, *IEEE J. Quantum Electron.* **QE-29**, 6, pp. 1817-1823.

Toyada, K., Miura, A., Urabe, S., Hayasaka, K. and Watanabe, M. (2001). Laser cooling of calcium ions by use of ultraviolet laser diodes: significant induction of electron-shelving transition, *Opt. Lett.* **26**, 23, pp. 1897-1899.

Troger, T., Thevenaz, L. and Robert, P. (1999). Frequency-sweep generation resonant self-injection locking, *Opt. Lett.* **24**, pp. 1493-1495.

Tsang, W. T., Olsson, N. A., Linke, R. A. and Logan, R. A. (1983). 1.5 μm wavelength GaInAsP C lasers: single-frequency operation and wide-band frequency tuning, *Electron. Lett.* **19**, pp. 415-416.

Tsang, W. T. (1985). The cleaved coupled-cavity (C^3) laser in *Semiconductors and semi-metals,* **22**, Part B, Academic Press, Orlando.

Tsuchida, H. (1994). Tunable, narrow-linewideth output from an injection-locked high-power A1GaAs laser diode array, *Opt. Lett.* **19**, 21, pp. 1741-1743.

Turner, L. D., Weber, K. P., Hawthorn, C. J. and Scholten, R. E. (2002). Frequency noise characterization of narrow linewidth diode lasers, *Opt. Commun.* **201**, pp. 391-397.

Uchiyama, Y. and Tsuchiya, M. (1999). Generation of ultraviolet(335 nm) light by intracavity frequency doubling from active mode-locking action of an external-cavity AlGaInP diode laser, *Opt. Lett.* **24**, 16, pp. 1148-1150.

Uemukai, M., Suhara, T., Yutani, K., Shimada, N., Fukumoto, Y., Nishihara, H. and Larsson, A. (2000). Tunable external cavity semiconductor laser using monolithically integrated tapered amplifier and grating coupler for collimation, *IEEE Photo. Tech. Lett.* **12**, 12, pp. 1607-1609.

Uetake, S., *XVIII International Conference on Atomic Physics,* Cambridge, Massachusetts, USA, pp. 43.

Vakhshoori, D., Tayebati, P., Lu, C. C., Azimi, M., Wang, P., Zhou, J. H. and Canoglu, E. (1999). 2 mW CW singlemode operation of a tunable 1550 nm vertical cavity surface emitting laser with 50 nm tuning range, *Electron, Lett.* **35**, 11, pp. 900-901.

Vanier, J. and Audoin, C. (1989). *The quantum physics of atomic frequency standards*, Adam Hilger, Bristol.

Varangis, P. M., Li, H., Liu, G. T., Newell, T. C., Stintz, A., Fuchs, B., Malloy, K. J. and Lester, L. F. (2000). Low threshold quantum dot laser with 201 nm tuning range, *Electron. Lett.* **36**, pp. 18-20.

Voumard, C., Salathe, R. and Weber, H. (1977). *Appl. Phys.* **12**, pp. 369.

Vukicevic, N., Zibrov, A. S., Hollberg, L., Walls, F. L., Kitching, J. and Robinson, H. G. (2000). *IEEE Trans. Ultrason. Ferroel.* **47**, pp. 1122.

Wacogne, B., Goedgebuer, J. P., Onokhov, A. P. and Tomilin, M. (1993). Wavelength tuning of a semiconductor-laser using nematic liquid crystals, *IEEE J. Quantum Electron.* **QE-29**, 4, pp. 1015-1017.

Wacogne, B., Goedgebuer, J. P. and Porte, H. (1994). Single lithium niobate crystal for mode selection and phase modulation in a tunable extended-cavity laser diode, *Opt. Lett.* **19**, 17, pp. 1334-1336.

Wandt, D., Laschek, M., Tunnermann, A. and Welling, H. (1997). Continuous tunable external-cavity diode laser with a double-grating arrangement, *Opt. Lett.* **22**, 6, pp. 390-392.

Wang, S. (1974). Priciples of distributed feedback and distributed Bragg reflector waveguides, *IEEE J. Quantum Electron.* **QE-10**, pp. 413.

Wang, C. L. and Pan, C. L. (1994). Tunable dual-wavelength operaition of a diode laser array with an external grating-loaded cavity, *Appl. Phys. Lett.* **64**, 23, pp. 3089-3091.

Wang, W. Z., Fejer, M. M. Hammond, R. H., Beasley, M. R., Ahn, C. H., Bortz, M. L. and Day, T. (1994). Atomic absorption monitor for deposition process control of aluminum at 394 nm using frequency-doubled didoe laser, *Appl. Phys. Lett.* **68**, 6, pp. 729-731.

Weidemuller, M., Gabbanini, C., Hare, J., Gross, M. and Haroche, S. (1993). A beam of laser-cooled lithium Rydberg atoms for precision microwave spectroscopy, *Opt. Commun.* **101**, pp. 342-346.

Welford, D. and Mooradian, A. (1982). Output power and temperature dependence of the linewidth of single-frequency CW (GaAl)As diode lasers, *Appl. Phys. Lett.* **40**, 10, pp. 865-867.

Wenzel, H., Klehr, A., Braun, M., Bugge, F., Erbert, G., Fricke, J., Knauer, A., Weyers, M. and Trankle, G. (2004). High-power 783 nm distributed-feedback laser, *Electron. Lett.* **40**, 2, pp. 123-124.

Westbrook, L. D., Nelson, A. W., Fiddyment, P. J. and Collins, J. B. (1984). Monolithic 1.5 μm hybrid DFB/DBR lasers with 5 nm tuning range, *Electron. Lett.* **20**, pp. 957-958.

Wicht, A., Rudolf, M., Huke, P., Rinkleff, R. H. and Danzmann, K. (2004). Grating enhanced external cavity diode lasers, *Appl. Phys. B* **78**, pp. 137-144.

Wieman, C. E. and Gillbert, S. L. (1982). Laser frequency-stabilization using mode interference from a reflecting reference interferometer, *Opt. Lett.* **7**, pp. 480-482.

Wieman C. E. and Hollberg, L. (1991). Using diode lasers for atomic physics, *Rev. Sci Instrum.* **62**, 1, pp. 1-20.

Williams, J. (1977). Electronic design, (1977).

Williamson, C. A., Adams, M. J., Ellis, A. D. and Borghesani, A. (2003). Mode locking of semiconductor laser with curved waveguide and passive mode expander, *Appl. Phys. Lett.* **82**, 3, pp. 322-324.

Wineland, D. J. and Dehmelt, H. G. (1975). Electrostatic frequency-shifts and broadening in a Penning trap, *Bull. Am. Phys. Soc.* **20**, pp. 637.

Woll, D., Schumacher, J., Roberson, A., Tremonet, M. A., Wallenstein, R., Katz, M., Eger, D. and Englander, A. (2002). 250 mW of coherent blue 460 nm light generated by single-pass frequency doubling of the output of a mode-locked high-power diode laser in periodically poled KTP, *Opt. Lett.* **27**, 12, pp. 1055-1057.

Woodward, S. L., Koren, U., Miller, B. I., Young, M. G., Newkirk, M. A. and Burrus, C. A. (1992). A DBR laser tunable by resistive heating, *IEEE Photo. Tech. Lett.* **4**, pp. 1330-1332.

Wu, I. F., Riant, F., Verdiell, J. M. and Dagenais, M. (1992). Real-time in situ monitoring of antireflection coating for semiconductor laser amplifiers by ellipsometry, *IEEE Photo. Tech. Lett.* **4**, 9, pp. 991-993.

Wu., M. S., Vail, E. C., Li, G. S., Yuen, W. and Chang-Hasnain, C. J. (1995). Tunbale micromachined vertical cavity surface emitting laser, *Elecron. Lett.* **31**, 19, pp. 1671-1672.

Wyatt, R. and Devlin, W. J. (1983). 10 kHz linewidth 1.5μm InGaAsP external cavity laser with 55 nm tuning range, *Eletron. Lett.* **19**, 3, pp. 110-112.

Yamamoto, Y. (1980). Receiver performance evaluation of various digital optical modulation-demodulation systems in the 0.5-1.0 μm wavelength region, *IEEE J.Quantum Electron.* **QE-16**, pp. 1251-1259.

Yamamoto, Y. and Kimura, T. (1981). Coherent optical fiber transmission systems, IEEE J. Quantum Electron. **QE-17**, pp. 919-935.

Yamamoto, Y. (1991). *Coherent amplification, and quantum effects in semiconductor lasers*, Chichester U.K., Wiley.

Yariv, A. (1991). *Optical electronics*, 4th Edition, Saunders College, Philadelphia.

Yashchuk, V. V., Budker, D. and Davis, J. R. (2000). Laser frequency stabilization using linear magneto-optics, *Rev. Sci. Instrum.* **71**, 2, pp. 341-346.

Ye, C. Y. (2004). Observation of electromagnetically induced transparency in Rb vapor by mode locked external cavity diode lasers, *Chin. Phys. Lett.* **21**, 5, pp.853-855.

Yilmaz, T., Depriest, C. M., Turpin, T., Abeles, J. H., and Delfyett, Jr., P. J. (2002). Toward a photonic arbitrary waveform generator using a modelocked external cavity semiconductor laser, *IEEE Photo. Tech. Lett.* **14**, 11, pp. 1608-1611.

Yokouchi, N., Miyamoto, T., Uchida, T., Inaba, Y., Koyama, F. and Iga, K. (1992). 40 A continuos tuning of a GaAsP/InP vertical-cavity surface-emitting laser using an external mirror, *IEEE Photo Tech. Lett.* **4**, 7, pp. 701-703.

Young, M. G., Koren, U., Miller, B. I., Chien, M., Koch, T. L., Tennent, D. M., Feder, K., Dreyer K. and Raybon, G. (1995). Six wavelength laser array with integrated amplifier and modulator, *Eletron. Lett.* **31**, 21, pp. 1835-1836.

Zimmermann, C., Vuletic, V., Hemmerich, A. and Hansch, T. W. (1995). All solid state laser source for tunable blue and ultraviolet radiation, *Appl. Phys. Lett.* **66**, 18, pp. 2318-2320.

Zimmermann, J., Struckmeier, J., Hofmann, M. R. and Meyn, J. (2002). Tunable blue laser based on intracavity frequency doubling with a fan-structured periodically poled $LiTao_3$ crystal, *Opt. Lett.* **27**, 8, pp. 604-606.

Zorabedian, P. and Trutna, W. R. (1990). Alignment-stabilized grating-tuned external-cavity semiconductor lasers, *Opt. Lett.* **15**, 9, pp. 483-485.

Zorabedian, P. (1992). Characteristics of a grating-external-cavity semiconductor lasers containing intracavity prism beam expanders, *J. Lightwave Technol.* **10**, pp. 330.

Zorabedian, P. (1994). Axial-mode instibility in tunable external-cavity semiconductor lasers, *IEEE J. Quantum Electron.* **QE-30**, 7, pp. 1542-1552.

Zorabedian, P. (1996). Tunable External-Cavity Semiconductor Lasers, pp. 349-442. Edited by F. J. Duarte, Eastman Kodak Company.

Zory, Jr., P. S. (1993). *Quantum well lasers*, Ed. Academic Press, San Diego.

Index

Printed in the United States
By Bookmasters